CATALYSTS FOR CHANGE

Statistical Thinking

A Simulation Approach to Modeling

Uncertainty

CATALYST PRESS

Copyright © 2015 Catalysts for Change

PUBLISHED BY CATALYST PRESS

This work is licensed under a Creative Commons Attribution 3.0 Unported License. You are free to share, remix, and to make commercial use of the work under the condition that you provide proper attribution. To reference this work, use

Zieffler, A., & Catalysts for Change. (2015). *Statistical Thinking: A simulation approach to uncertainty* (third edition). Minneapolis, MN: Catalyst Press.

The work to create the material appearing in the book was made possible by the National Science Foundation (DUE–0814433).

Printed in the United States of America

ISBN 978-0615691305

Catalyst Press
Minneapolis, MN 55455

http://catalystsumn.blogspot.com

Third printing, August 2015

COLOPHON

The primary font used in the book is Libre Baskerville. Headings are set in Neutra Display, and TinkerPlots™ commands are printed using Linux Libertine Mono.

Graphics in the book (not listed below) were in the public domain and were obtained from:

- Course activity and reading icons from [Open Clip Art Library](#).
- Dog icons used in the Matching Dogs to Owners visualization by Sara Quintana from the [Noun Project](#). (CC License)
- TinkerPlots™ icons and screenshots are used by permission from [Key Curriculum Press](#).

Colophon	4
Introduction	11
Graphical Icons	12
TinkerPlotsTM Software	12
Data and Other Resources for the Book	13
Mac Users	13
Participation in the Learning Process	13
Unit I: Modeling & Simulation	15
Simulation	16
Policy and Population	17
Outline of the Unit	20
Course Activity: Can You Beat Randomness	21
Randomness	24
Course Activity: iPod Shuffle	25
Group Task	26
Explore and Describe	26
Develop Rules	27
Test Rules	27
Evaluate	28
Summarize	28
Discussion	28
Probability Simulation	30
Course Activity: Modeling Random Behavior—Coin Flips	31
Intuitions about Coin Flips	32
Modeling Coin Flips	33
Collecting the Results from Many Trials	37
Plotting the Results from the Trials	38
Course Activity: Modeling Random Behavior—Dice Rolls	41
Modeling Dice Rolls	42
Extensions	45
Course Activity: Automating the Summarization and Collection of Simulation Results	47

Modeling Coin Flips	47
Automating the Collection of Trial Results	48
Modeling Dice Rolls	53
Extensions	55
Introduction to Statistical Hypothesis Testing	**56**
Models and Hypotheses	57
Connections	59
Helper or Hinderer Readiness	**60**
Course Activity: Helper or Hinderer	**61**
Discuss the Following Questions	62
The "Just-by-Chance" (No Preference) Model	63
Summary of the Simulation Process	64
Simulating the Data	65
Evaluating the Hypothesized Model	66
Quantifying the Level of Support For a Model Based on the Observed Result	67
Mission Improbable: A Concise and Precise Definition of p-value	**72**
Course Activity: Monday Breakups	**73**
Discuss the Following Questions	74
Simulating the Data	75
Evaluating the Hypothesized Model	76
Course Activity: Racial Disparities in Police Stops	**79**
Discuss the Following Questions	80
Simulating the Data	82
Evaluating the Hypothesized Model	83
Learning Goals: Unit 1	**85**
Literacy/Understanding (Terms and Concepts)	85
Model–Simulate–Evaluate Framework	86
TinkerPlots™ Skills	86
Course Activity: Unit 1 Wrap-up & Review	**87**
Terminology for Unit 1	87
Modeling Real-World Phenomena	88

    p-Value   89

    Further Practice   92

Unit II: Comparing Groups   95

    Comparisons in the Media   96

    Statistical Comparisons   98

    Operationalization   98

    Summarization   100

    To Infer or Not to Infer   103

    Outline of the Unit   104

    References   105

America's Most Reliable Airlines   106

Course Activity: Arrival Delay Times   108

    Group Task   109

    Explore and Describe   109

    Develop Rules   110

    Test Rules   111

    Summarize   111

    Discussion   111

Course Activity: Memorization   113

    Examining the Observed Data}   114

    Summarizing the Difference Between the Two Conditions   115

    Considering Chance Variation as an Explanation for the Difference in Means   116

    The "Just-by-Chance" Model   117

    Re-randomization: Inspecting Other Possible Random Assignments of the Subjects   118

    Examining the Distribution of the Difference in Means   120

Sleep Deprivation Readiness   123

Course Activity: Sleep Deprivation   124

    Modeling the Differences in Improvement Under the "just-by-chance" Model   128

    Randomization Tests in TinkerPlotsTM   128

    Collecting and Plotting the Difference in Means   135

Evaluating the Hypothesized Model	136
Example Write-Up for Sleep Deprivation Study	137
Course Activity: Latino Achievement	**139**
Discuss the following questions.	140
Examine the Data	140
Modeling Achievement Under the "Just-by-Chance" Model	142
Plotting and Collecting the Results	142
Simulate Results under the "Just-by-chance" model	143
Evaluating the Hypothesized Model	144
Random Assignment	**146**
Course Activity: Strength Shoe®	**147**
Discuss the following questions.	147
Random Assignment	149
Confounding Variables	150
Observed Variable: Sex	151
Observed Variable: Height	154
Unobserved Confounding Variables	156
Random Selection	**159**
How are Study Participants Assigned to Groups/Conditions?	159
How are Study Participants Initially Selected to be a Part of the Study?	160
Studies that Use Random Sampling	160
Method of Analysis	161
Course Activity: Sampling	**162**
Unbiasedness	164
Simple Random Sampling	169
Sample Size	172
Population Size	174
Course Activity: Dolphin Therapy	**177**
Discuss the following questions.	178
Modeling the Improvement Under the "Just-by-Chance" Model	180
Plotting and Collecting the Results	181
Simulate Results under the "Just-by-chance" model	182

Evaluating the Hypothesized Model	183
Observational Studies	185
Course Activity: Murderous Nurse	187
Discuss the Following Questions	188
Modeling the Chance Variation Under the Assumption of No Difference	190
Plotting and Collecting the Results	190
Simulate Results under the "Just-by-chance" model	190
Evaluating the Hypothesized Model	192
Learning Goals: Unit 2	193
Literacy/Understanding (Terms and Concepts)	193
Model–Simulate–Evaluate Framework	194
TinkerPlotsTM Skills	194
Course Activity: Unit 2 Wrap-up & Review	195
Terminology for Unit 2	195
Rating Chain Restaurants	198
Native Californians	199
Blood Pressure	201
Social Fibbing	203
Unit III: Statistical Estimation	205
Outline of the Unit	206
References	206
Describing Distributions	207
Shape	208
Location	209
Variation	210
Putting It All Together	211
Course Activity: Features of Distributions	212
Cell Phone Bills	215
Number of Hours Studied	216
Understanding the Standard Deviation	218
Course Activity: Comparing Hand Spans	219
The Standard Deviation	220

> Using Both the Mean and Standard Deviation: A Complete Summary — 223

Course Activity: Kissing the 'Right' Way — 225
- Discuss the following questions. — 226
- The Standard Error — 227
- Uncertainty of the Estimate — 228
- Modeling the Variation Due to Random Sampling — 228
- Margin of Error — 231
- Interval Estimates — 232

Margin of Error — 233

Course Activity: Memorization (Part II) — 234
- Effect Size — 235
- Interval Estimate for the Effect Size — 241

Course Activity: Why Two Standard Errors — 242
- Exploring Random Samples from an Unknown Population — 242
- Exploring a Single Random Sample — 245
- Exploring the Bootstrap Distribution from Different Random Samples — 246
- Exploring the Bootstrap Intervals Computed from Different Random Samples — 247
- What does it Mean to be 95% Confident? — 248

Learning Goals: Unit 3 — 251
- Literacy/Understanding (Terms and Concepts) — 251
- Selecting/Using Models — 252
- TinkerPlotsTM Skills — 252

Course Activity: Unit 3 Wrap-up & Review — 253
- Terminology for Unit 3 — 253
- Relationship Between p-Value and Standard Error — 254
- Balsa Wood — 256
- MicrosortTM — 257
- Marijuana — 259

INTRODUCTION

Learning statistics is sexy. Hal Varian, Google's chief economist, believes this. During an interview in McKinsey Quarterly, Varian stated, "I keep saying the sexy job in the next ten years will be statisticians. People think I'm joking, but who would've guessed that computer engineers would've been the sexy job of the 1990s?" Varian is not the only person to express this sentiment either. Hans Rosling in the 2010 BBC documentary *Joy of Stats*[1] referred to statistics as the "sexiest subject around".

Whether you believe it is the sexiest subject or not, it is incontrovertible that the use of statistics and data are prevalent in today's information age. Almost every person on earth will benefit from learning some foundational ideas of statistics. This is true because statistics forms the basis of our everyday world just as much as do science, technology, and politics. Google, Netflix, Twitter, Facebook, OKCupid, Match.com, Amazon, iTunes, and the Federal Government are just a handful of the companies and organizations that use statistics on a daily basis. Journalism, political science, biology, sociology, psychology, graphic design, economics, sports science, and dance are all disciplines that have made use of statistical methodology.

The materials in this book will introduce you to the seminal ideas underlying the discipline of statistics. In addition, they have been designed with your learning in mind. For example, many of the in-class activities were developed using pedagogical principles, such as small group activities and discussion, that have been shown in research to improve student learning.

---

[1] Watch *Joy of Stats* online at http://www.gapminder.org/videos/the-joy-of-stats/

GRAPHICAL ICONS

The use of graphical icons throughout the text are intended to help you make your way through the material. Throughout the book we denote each section as an in-class activity or course reading. The graphical icon to denote these are:

 In-class Activity

 Course Reading

The course readings should be completed outside of class and are intended to help you learn and extend the ideas, skills, and concepts you learn in the classroom. The readings themselves are not all "traditional" readings in the sense of words written on paper, but instead often link to video clips, blogs and other multimedia material.

TINKERPLOTS™ SOFTWARE

Much of the material presented in the book requires the use of TinkerPlots™. This software can be purchased and downloaded (for Mac or PC) from http://www.tinkerplots.com/.

DATA AND OTHER RESOURCES FOR THE BOOK

The data sets used in the materials, as well as other materials that accompany the book are available at http://z.umn.edu/statisticalthinking. Clicking this link will download a ZIP file to your computer. Double-click on the ZIP file to view all the materials.

MAC USERS

If you are using a Mac and seem to have problems downloading the TinkerPlots™ data files, hold the option-key while clicking on the link. This should download the file to your desktop (or your *Downloads* folder). You then need to erase the .txt suffix that is appended to the end of the downloaded file in order to use the file with TinkerPlots™. The file suffix should be .tp, and not .tp.txt.

PARTICIPATION IN THE LEARNING PROCESS

The textbook, instructors, and teaching assistants are all resources that are at your disposal to help you learn the material. In the end, however, you will have to do all of the hard work associated with actually learning that material. To successfully navigate this process, it is vital that you be an active participant in the learning process. Coming to class, participating in the activities and discussions, reading, completing the assignments, and asking questions are essential to successful learning.

Learning anything new takes time and effort and this is especially true of learning statistics, as you are not just learning a set of methods, but rather a disciplined way of thinking about the world. Changing your habits of mind will take continual practice. It will also take a great deal of patience and persistence.

As you engage in and use the skills, concepts and ideas introduced in the material, you will find yourself thinking about data and evidence in a different way. This may lead you to make different decisions or choices. But, even if this course does not change your world overnight, you will at the very least be able to critically think about inferences and conclusions drawn from data.

UNIT I: MODELING & SIMULATION

There is mounting evidence that the "model-building era" that dominated the theoretical activities of the sciences for a long time is about to be succeeded or at least lastingly supplemented by the "simulation era".

—S. Hartmann (1996)

Modeling is one of the most important subjects you may ever learn. It is used in microbiology, macroeconomics, urban studies, sociology, psychology, public health, computer science, and of course, statistics. In fact, modeling is a method that is used in almost every discipline. Many think that it is an important skill to learn because it is so pervasive. While this is true, even more important is how closely the skills of modeling tie to the more general skills of problem solving. Starfield, Smith, and Bleloch (1994) summed this sentiment up nicely when they wrote, "learning to model is bound up with learning to solve problems and to think imaginatively and purposefully" (p. x)[2].

A model is a simplified representation of a system that can be used to promote an understanding of a more complex system. For example, meteorologists use computers to build models of the climate to understand and predict the weather. The computer model includes behaviors or properties which correspond, in some way, to the particular real-world system of climate. The computer models, however, do not include every possible detail about climate. All models leave things out and get some things—many

---

[2] Starfield, A. M., Smith, K. A., & Bleloch, A. L. (1994). *How to model it: Problem solving for the computer age.* Edina, MN: Burgess International Group, Inc.

things—wrong. This is because all models are simplifications of reality. Since all models are simplifications of reality there is always a trade-off as to what level of detail is included in the model. If too little detail is included in the model one runs the risk of missing relevant interactions and the resultant model does not promote understanding. If too much detail is included in the model, the model may become overly complicated and actually preclude the development of understanding.

Models have many purposes[3], but are primarily used to better understand phenomena in the real-world. Common uses of models are for description, exploration, prediction, and classification. For example, Google builds models to understand and predict peoples' internet searching tendencies. These models are then used to help Google carry out more efficient and better searches of information. As another example, Netflix builds models to understand the characteristics of movies that their customers have rated highly so that they can then recommend other movies that the person may enjoy. Amazon and Apple iTunes both use models in similar manners.

SIMULATION

One method that statisticians use to understand real-world phenomena is to conduct a simulation. A simulation is the manipulation of a model to enable a person to understand the potential outcomes and interactions of the system being modeled. In a simulation, a model is used to generate data under a particular set of conditions or assumptions. These conditions and assumptions allow the model to mimic processes and events in the real-world. By examining the data produced from the simulation, researchers can draw insight about and predict what might happen in the real-world under a given set of circumstances. Consider the following example:

---

[3] Joshua Epstein in his keynote address Why Model? offers several additional purposes for building models.

POLICY AND POPULATION

In 1978, China introduced the "one-child" policy in order to alleviate social, economic, and environmental problems in China. According to Wikipedia[4],

> The policy officially restricts the number of children married urban couples can have to one, although it allows exemptions for several cases, including rural couples, ethnic minorities, and parents without any siblings themselves. A spokesperson of the Committee on the One-Child Policy has said that approximately 35.9% of China's population is currently subject to the one-child restriction.

Although the Chinese government has suggested that the policy has prevented more than 250 million births from its implementation to 2000, the policy is controversial both within and outside of China because of the manner in which the policy has been implemented. There have also been concerns raised about potential negative economic and social consequences, in part because many families were determined to have a son. Scholars have wondered how things would change if instead of a one-child policy, a country adopted a "one son" policy. A "one son" policy would allow families to keep having children until they had a son. If a family's first child is a boy, they would be restricted from having more children. If, however, the first child was a daughter, the family could continue having children until a son was born.

If the United States adopted a "one son" policy, how would the policy affect the average number of children per family, which is currently 1.86?

---

[4] One-child policy. (2015, May 30). In *Wikipedia, The Free Encyclopedia*. Retrieved 18:02, June 1, 2015, from http://en.wikipedia.org/w/index.php?title=One-child_policy&oldid=664745432

One way in which this question could be studied (without actually implementing the policy) would be to conduct a simulation study by modeling this situation. Consider for a minute how you might model the number of children a particular family would have.

One way to model this is to write the word **boy** on one index card and the word **girl** on another index card and to place those two index cards in a hat. After mixing up the index cards, you could draw a single card from the hat. If the card has the word **boy** written on it, the simulated "family" would be reported to have one child. If the card has the word **girl** written on it, a tally mark could be recorded and the index card would be replaced in the hat. The cards could then be remixed and another card would be drawn. If the card drawn has the word **boy** written on it, the simulated "family" would be reported to have two children. If the card has the word **girl** written on it, another tally mark could be recorded and the index card would again be replaced in the hat. This process would continue until the **boy** card was drawn. The table below shows the results after carrying out this process for three simulated families.

The recorded number of girls and boys for three simulated families.

Family	Girl	Boy
Family #1	✓	✓
Family #2		✓
Family #3	✓✓	✓

We could carry out this simulation for many families, say 500 families, and use the results to provide an answer to the research question. You can imagine that carrying out even this simple simulation would quickly become quite tedious. Simulation studies, such as this, are typically carried out using computer programs. In this unit, you will learn to use a computer program called TinkerPlots™ to model processes in the real-world and carry out simulation studies.

"Wait," you say. "Even if I carried out this simulation, I still would not be able to provide an answer to the research question! It doesn't reflect reality! Some families may not want to have any children, while others might be happy to stop after a girl was born. What about multiple births?" Maybe you are even questioning whether the probability of having a boy or having a girl is really 50:50. These are all valid points, and all would likely affect the results of the simulation, which in turn affects the inferences and conclusions that are drawn.

Often incredibly complex models are used in carrying out research. As an example, Electronic Arts, the video game company behind titles such as *Madden*, *NHL* and *FIFA*, uses game telemetry (the transmission of data from a game executable for recording and analysis) to model the gameplay patterns of players and identify the elements of their games that are highly correlated with player retention[5]. By understanding the behavior of players and the common patterns that are used, Electronic Arts game developers can focus their attention on more relevant features in future iterations of the game and ultimately reduce production costs. In their examination of *Madden NFL 11*, Electronic Arts used 46 features to model a player's mode preferences, control usage, performance, and play-calling style. This is but one example of using simulation in video games. To see other applications of how data are being used in video game design, watch the webinar, How Big Data and Statistical Modeling are Changing Video Games.

While the model we used in the previous example is overly simplistic for drawing any sorts of conclusions about implementing a "one son" policy in the United States, it could however, provide a useful starting point for introducing additional complexity. Even in the most enormously complicated modeling problem, researchers often make many simplifying assumptions[6]. With enough simplification, a model can be constructed and studied. The model is evaluated and often revised or updated as certain assumptions are deemed tenable and others are not. Because of this process, simulation studies are generally iterative in their development. This iteration process continues until an adequate level of understanding is developed and the research question can be answered.

---

[5] Weber, B. G., John, M., Mateas, M., & Jhala, A. (2011). Modeling player retention in Madden NFL 11. Presented at *Innovative Applications of Artificial Intelligence.* http://users.soe.ucsc.edu/~bweber/pubs/madden11retention.pdf

[6] Remember that *all* models—even those that seem quite complex—are simplifications of reality and get many things wrong.

OUTLINE OF THE UNIT

In this unit, you will begin by exploring ideas related to randomness. Randomness permeates, and is, in fact, fundamental to understanding the models that are used in statistics. In learning about these ideas, you will also be confronted with common human intuitions about randomness that are incorrect and can be misleading.

After examining the 'behavior' of randomness, you will be introduced to the TinkerPlots™ software, and learn how to model simple random processes such as coin flips and dice rolls. You will also learn how to carry out a simulation using TinkerPlots™. At the end of this unit, you will learn how the models introduced in this unit are used to examine research hypotheses about the world. In particular, you will use a chance model to evaluate results that have been observed in research studies to judge the level of evidence for a particular tested hypothesis.

As you progress through the unit, remember that the modeling process is a creative process that can often be very challenging. At times, this might lead to frustration as you are learning and practicing some of the material. But, as Mosteller et al. (1973) remind us, it is also a profitable experience since "modeling is not only a technique of statistics…it is a method of study which can be applied in many other fields as well" (p. xii)[7].

---

[7] Mosteller, F., Kruskal, W. H., Link, R. F., Pieters, R. S., & Rising, G. R. (1973). *Statistics by example: Finding models*. Reading, MA: Addison–Wesley.

COURSE ACTIVITY: CAN YOU BEAT RANDOMNESS

In his best-selling book The Drunkard's Walk: How Randomness Rules Our Lives, Leonard Mlodinow describes a probability guessing game that psychologists have conducted with different species such as humans, rats, and other non-human animals: Subjects are shown a sequence of red and green lights or cards, in which the colors appear randomly with different probabilities but otherwise with no pattern.[8] The probability of a specific color stays the same. After subjects watch the colors appear for a while, they are then asked to predict the color that will appear on subsequent lights or cards.

You are going to play a game, similar to the one described by Mlodinow.

- You will first watch a sequence of colors at: https://github.com/zief0002/Statistical-Thinking/blob/master/images/animation.gif. (Note: Although the sequence is only 50 generated squares, it will repeat, so stop the animation when you are satisfied.)
- After watching the sequence of colors, access the guessing game at http://statisfactions.shinyapps.io/guessing-game/
  - Based on the sequence you just watched, predict the color of the next light. To make a prediction, click once on the radio button for "Red" or "Green".
  - Once you make your prediction, the program will display your prediction in the "Your Guess" row. The program will then randomly generate a color and display it in the "Actual Result" row.

---

[8] Mlodinow, L. (2008). *The drunkard's walk: How randomness rules our lives.* New York: Pantheon.

- On the top of the window, you can also see the total number of predictions you have made, how many of those predictions were correct (matched the outcome to the actual color), as well as the percentage of your total predictions that were correct.

1. Play the game one time (i.e., make 50 predictions). What percentage of your 50 predictions were correct?

2. Describe the strategy that you used to make your predictions.

3. Play the game again using the strategy you just described. What percentage of your 50 predictions were correct this time?

4. Suppose that while playing the game you observed 10 green lights in a row. Can you predict the color of the **next light**? Explain.

5. Suppose that while playing the game you observed 20 green lights in a row. Can you predict the color of the **next light**? Explain

RANDOMNESS

To follow up on the *Can You Beat Randomness* activity, we will continue the exploration of randomness and human beings' intuitions of randomness. In upcoming class periods, you will explore some of your own intuitions of probabilistic devices. To prepare for this, you should do the following:

- Watch the YouTube video *Random Sequences: Human vs. Coin*
  http://www.youtube.com/v/H2lJLXS3AYM

- Read a blog entry about randomness called *What is Randomness?*
  http://sxxz.blogspot.com/2005/08/what-is-randomness.html

- Watch the YouTube video *Fun Science: Randomness*
  http://www.youtube.com/watch?v=t8mSHosc0Dk

COURSE ACTIVITY: IPOD SHUFFLE

Share and discuss your responses to each of the following questions with your group.

6. Do you have an iPod or some other digital music player? Have you used the shuffle feature? If you have used the shuffle feature, have you ever wondered how truly random it is?

7. What comes to mind when you hear the word, 'random'?

8. If the iPod shuffle feature is not producing a random sequence of songs, then what might the sequence of songs look like? What would you expect to see?

9. Do you think you can be 100% certain that a sequence of songs was not randomly generated? Explain your answer.

GROUP TASK

Albert Hoffman, an iPod owner, has written a letter to Apple to complain about the iPod shuffle feature. He writes that every day he takes an hour-long walk and listens to his iPod using the shuffle feature. **He believes that the shuffle feature is producing playlists in which some artists are played too often and others are not played enough.**

He has claimed that the iPod Shuffle feature is not generating random playlists. As evidence, Mr. Hoffman has provided both his music library (8 artists with 10 songs each) and three playlists (20 songs each) that his iPod generated using the shuffle feature.

Tim Cook, the CEO of Apple, Inc., has contacted your group to respond to Mr. Hoffman's complaint. He has provided your group with several playlists of 20 songs each using the same songs as Mr. Hoffman's library but generating them using a genuine random number generation method.

To help your group respond to Mr. Hoffman, the next four sections of the problem are designed to help your group explore properties of the randomly generated lists to develop rules that could help determine whether a set of playlists provide evidence that the shuffle feature is not producing randomly selected songs.

EXPLORE AND DESCRIBE

Examine the randomly generated playlists (your group will be given 25) to get an idea of the characteristics of these lists. Write down at least two characteristics about the randomly generated playlist that help you address Mr. Hoffman's concern.

DEVELOP RULES

Use the set of characteristics that your group wrote down to describe randomly generated playlists in the previous section to create a set of one or more rules that flag playlists that **do not appear to have been randomly generated**. (Be sure that each of the characteristics in the previous section is included in a rule.) *These rules should be clearly stated so that another person could easily use them.*

TEST RULES

Your group will be given five additional randomly generated playlists on which to test your rules. Let your instructor know that you are ready to receive these playlists. See whether the set of rules your group generated would lead someone to (incorrectly) question whether these playlists are not randomly generated. Based on the performance of your group's set of rules, adapt or change the rules as your group feels necessary.

EVALUATE

Your group will be provided with Mr. Hoffman's original three playlists. Apply your group's rules to these three playlists to judge whether there is convincing evidence that Mr. Hoffman's iPod Shuffle feature is producing playlists which do **not** seem to be randomly generated.

SUMMARIZE

Your group will now write a letter to Mr. Hoffman that includes the following:

- Your group's set of rules, used to judge whether a playlist does not appear to have been randomly generated. In your letter the rules need to be clearly stated so that another person could apply them to a playlist of 20 songs from Mr. Hoffman's music library;
- A response to Mr. Hoffman's claim that the shuffle feature is **not random** *because it produces playlists in which some artists are played too often and others are not played enough.*

Type the letter in a word-processed document and email it to each of your group members and the instructor.

DISCUSSION

As a group, discuss your responses to each of the following questions.

10. What made it difficult to come up with a rule to determine whether a sequence of data had been randomly generated? Explain.

11. How might your rules *change* if there were **not an equal number of songs for each artist**? Or a longer set of songs per playlist? (Be specific about how your rules might change.)

12. What does your group need to do to improve the process of working as a team? Be specific about how each member of the group will contribute to this improvement.

PROBABILITY SIMULATION

To prepare you for the next two course activities, *Modeling Random Behavior–Coin Flips* and *Modeling Random Behavior–Dice Rolls,* you will get a chance to practice creating models using TinkerPlots™. In upcoming class periods, you will add to these fundamental TinkerPlots™ skills to build increasingly complex models and carry out simulations. One helpful resource for learning TinkerPlots™ are the tutorial videos. These are available both online (at http://www.keycurriculum.com/PreBuilt/tp2/movies/) and via the Help menu in TinkerPlots™.

- Watch the TinkerPlots™ movie *Probability Simulation*. This will help to build some of the ideas that we will be working with in class. http://www.keycurriculum.com/PreBuilt/tp2/movies/probability-simulation.html

COURSE ACTIVITY: MODELING RANDOM BEHAVIOR—COIN FLIPS

In the book <u>Randomness</u>, author, Deborah J. Bennett[9], states,

> Everyone has been touched in some way by the laws of chance. From shuffling cards to start a game of bridge to tossing a coin at the start of a football game---most of humanity encounters chance daily. Our vocabulary is full of phrases that involve chance: likely; unlikely; probability; odds; chances; random; etc.

Bennett describes the concept of randomness as being deceptively complex because many aspects of it are counterintuitive. In particular, she points out that misconceptions about randomness and probability are dangerous due to the constant use of statistics, probabilities and odds in everyday life.

You will be exploring the following questions:

> How can we use TinkerPlots™ to model random outcomes of common chance devices such as coins and dice? How can we use the results from a simulation to evaluate our intuitions about the world?

---

[9] Bennett, D. J. (1998). *Randomness*. Cambridge, MA: Harvard University Press.

INTUITIONS ABOUT COIN FLIPS

Questions 1–7 are asking for your intuitions. You do not have to calculate exact values. We will explore these questions in more detail later in this activity.

1. Imagine that you flip a fair coin ten times and record the outcome for each flip. How many heads would you expect to see?

2. What is the smallest number of heads you could obtain?

3. What is the largest number of heads you could obtain?

Imagine flipping a fair coin ten times and recording the outcome for each flip. Once you obtain the ten flips, you count the number of heads. This process is repeated 100 times. Each time you record the number of heads out of the ten.

4. Which result(s) for the number of heads would you expect to occur the most often?

5. How often—what percentage of the 100 sets of 10 coins—would you expect to get a result of five heads?

6. How often—what percentage of the 100 sets of 10 coins—would you expect to get a result of all ten heads?

7. Which result, two heads or eight heads, would you expect to see more often? Why?

MODELING COIN FLIPS

To save time and to gather data quickly, you will use a software package called TinkerPlots™ to model tossing a coin 10 times. You will use the data you generate with the simulation to check your initial intuitions about what would happen if you repeatedly tossed a single coin 10 times.

In simulation studies, a model is used to generate data. A model is a representation for a particular purpose. Models can be used for description, exploration, prediction, or classification. A model might be a physical object or it might be an idea or construct.

- Set up the model for flipping a coin (see instructions below).
- After you have set up the model, click the Run button.
- A *case table* displaying the outcome for the "coin flip" should have been produced.

Plotting the Trials' Results
- Highlight the *Results* attribute from your case table.
- Drag a new `Plot` from the object toolbar into your document.

- All five trial results are now displayed in your plot, but in an unorganized manner.
- Grab one of the circle icons in the plot and drag it to the right. This will help organize the counts. (Note. You may want to resize the window so it is larger.)
- Keep dragging until you are satisfied that you can answer the questions posed earlier.
- Although the data are now more organized along the horizontal axis, the heights are still arbitrary. You can change this by clicking the `Stack Vertical` button in the upper plot toolbar.

Bin Lines

Each time you create a plot, vertical bars—called bin lines—will be drawn in by default. When you separate cases by dragging, TinkerPlots™ draws additional bin lines. These lines show how groups of cases are separated. Labels for each separated group (such as "0–4" and "5–9") are also added to the plot.

To separate cases so that each group only contains a single value, double-click on one of the endpoints and change the value for `Bin width` (i.e., `Bin width=1`) and click `OK`. To fully separate the cases so that there are no groups, drag a case icon until the bin lines disappear.

Setting Up the Simulation
- Drag a new `Sampler` from the object toolbar into your blank document.

- Drag a new `Spinner` from the device toolbar at the bottom of the `Sampler` onto the current device.

- There are two outcomes in our spinner, *a* and *b*. Click on each of these and re-label them *Heads* and *Tails*.
- From the `Device Options` menu (to the lower-left of the `Spinner`) select `Show Proportion`.

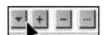

- In modeling a coin flip, the two outcomes *Heads* and *Tails* are equally likely events. Therefore you need to change the proportions so that each is 0.5. Change this by either dragging the dividing line in the spinner or by clicking on the proportion and changing it to the desired value.
- Change the `Draw` value from *2* to *1* and the `Repeat` value from *5* to *1*. This simulates tossing one coin (`Draw`) one time (`Repeat`).
- Change the `Spinner` label from *Attr1* to *Coin*.

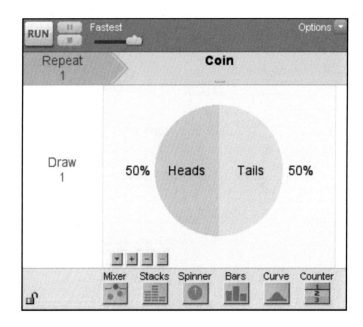

A TinkerPlots™ Sampler Showing the Model for Flipping a Single Coin Once

Record the result in the first cell of "Trial 1" in the table below. Run the simulation nine more times and record the outcome for each additional "flip" in the "Trial 1" column.

	Trial 1	Trial 2	Trial 3	Trial 4	Trial 5
Flip #1					
Flip #2					
Flip #3					
Flip #4					
Flip #5					
Flip #6					
Flip #7					
Flip #8					
Flip #9					
Flip #10					

8. How many *heads* did you get in the ten "flips" for Trial 1?

In simulation experiments, each time the model is used to produce outcomes, it is referred to as a *trial*. A trial can consist of one or many outcomes depending on the simulation. In this simulation, the trial consists of flipping the coin 10 times.

We are often more interested in some quantification from the trial rather than the individual outcomes that make up the trial. In this case, we are interested in the number of heads that are "tossed." This quantification, or *result*, of the trial is typically recorded and then examined along with the results from several other trials to understand a phenomenon under study.

9. Carry out four more trials of the simulation. Record the outcomes in the table above. Also record the result (i.e., the number of heads) from each trial.

COLLECTING THE RESULTS FROM MANY TRIALS

In order to answer the questions posed previously, you need to collect the results from many trials of the simulation experiment.

- Enter the results from each of your five trials into a TinkerPlots™ case table (see instructions on next page).

Collecting the results from each trial of a simulation into a case table allows us to organize and display the results from multiple trials. For example, right-clicking on the attribute name (if you are on a Mac press the control key while you click), brings up a menu from which you can sort these results in either ascending (Sort Rows Ascending) or descending (Sort Rows Descending) order. This will allow us to answer questions such as 'what is the maximum number of heads we would expect to see'.

> **Collecting the Results from Each Trial**
> - Drag a new `Case Table` from the object toolbar into your blank document
>
>
>
> - Click on `<new>` to change the attribute name. Rename this attribute *Results*.
> - Enter the results from each trial (the number of heads) into the results column in the case table.

PLOTTING THE RESULTS FROM THE TRIALS

Although you can see the result from each trial in the case table, this is not a good way to understand the results—especially when there are results from more than just a few trials. For example, consider trying to compare the number of trials in which you got *two heads* and the number of trials in which you got *eight heads*. **A better way to organize simulation results, after entering them into a case table, is to plot them.**

- Plot the results from your five trials (see instructions on next page).

10. Sketch the plot below.

Can you answer the questions posed earlier using your simulation results? You may not feel comfortable answering those questions based on the results from only five trials. You should carry out more trials and add the results to our case table.

- In the Sampler from which you originally ran the simulation, change the Repeat value from *1* to *10*.
- Click the Run button.

This simulates tossing one coin (Draw) 10 times (Repeat).

- Add the trial result to your case table of results. Notice that the plot of the results automatically updates when a result is added to the case table.
- Carry out several more trials of the simulation. Add each trial result into your case table. Be sure you have the results from *at least ten trials*.
- Enter the result from each of your ten trials into your instructor's computer.

11. After each group in the class has entered their data, sketch the plot of results shown on your instructor's computer. Be sure to add an appropriate scale and label to your x-axis so that you can respond to the questions below.

12. Which result(s) for the number of heads occurred the most often?

13. About what percentage of the trials resulted in five heads?

14. About what percentage of the trials resulted in ten heads?

15. Which result, two heads or eight heads, occurred more often?

16. Sketch the plot you would expect to see if you could simulate the results from 10,000 trials of the coin flipping experiment, but instead of flipping the coin ten times, you flipped it 20 times.

COURSE ACTIVITY: MODELING RANDOM BEHAVIOR—DICE ROLLS

Imagine that you are rolling a six-sided die ten times. In those ten rolls, consider the number of times that you would expect to see the outcome of three.

Imagine repeating this process 100 times.

Questions 1–4 are asking for your intuitions. You do not have to calculate exact values. We will explore these questions in more detail later in this activity.

1. Which outcome (0 threes to 10 threes) would you expect to occur most often?

2. What percentage of the time would you expect to see an outcome of five threes? Explain.

3. Which outcome, two threes or eight threes, would you expect to see more often? Why?

4. What percentage of the time would you expect to get an outcome of all ten threes?

MODELING DICE ROLLS

Now you will use TinkerPlots™ to simulate rolling a single die ten times. You can again use the data you generate in the simulation to check your initial intuitions about the number of threes that would occur in ten rolls.

- Open a new document in TinkerPlots™. To open a new document, select New from the File menu.
- Set up the model of tossing a single die ten times dice (see instructions on next page).
- After you have set up the model, click the Run button to carry out a single trial of the simulation.

Setting Up the Simulation
- Drag a new Sampler from the object toolbar into your blank document.
- This time rather than changing the Sampler device to a Spinner, you will use the default device of the Mixer.
- Click the Add Element button (the plus sign) until you have six elements in your mixer.

- Click on each element label and change the labels to the numbers *1* through *6* (so that each element represents one side of a die).
- Change the Draw value from *2* to *1* and the Repeat value from *5* to *10*. This simulates rolling a single die (Draw) ten times (Repeat).
- Change the mixer label from *Attr1* to *Dice*.

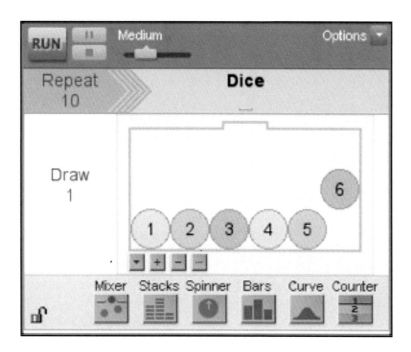

TinkerPlots™ Sampler Showing the Model for Rolling a Single Die Ten Times

In this simulation, the trial consists of rolling the die ten times. The result from each trial is the number of threes that occur.

- Record the number of threes that occurred in the trial in a case table. (If you've forgotten how to get a case table, go back and re-read the directions from the previous Course Activity.)
- Carry out nine more trials. Record the results from each trial into your case table.
- Enter the result from each of your ten trials into your instructor's computer.

Add the results from at least four other groups to your case table so that you have the results from 50 trials. Use the plot of the results from the 50 trials of the simulation to answer each of the following questions.

5. Create a plot of the results. (If you've forgotten how to plot the results, go back and re-read the directions from earlier in this activity.) Sketch the plot here. Be sure to label the x-axis.

6. Which result occurred most often?

7. What percentage of the time did a result of *five threes* occur?

8. Which result, *two threes* or *eight threes*, occurred more often?

9. What percentage of the time did a result of *ten threes* occur?

EXTENSIONS

10. Compare both plots of the results with those of classmates from at least two other groups. Are your plots identical? Comment on the similarities and differences.

11. Based on your examination of these plots, sketch the plot you would expect to see if you could simulate the results from 10,000 trials (rather than 50) of the die rolling experiment.

12. Sketch the plot you would expect to see if you could simulate the results from 10,000 trials of the die rolling experiment, but instead of rolling the die ten times, you rolled it 20 times.

COURSE ACTIVITY: AUTOMATING THE SUMMARIZATION AND COLLECTION OF SIMULATION RESULTS

In previous activities and assignments, you have learned how to set up a model to run a simulation experiment using TinkerPlots™. In these simulations, you ran many trials from which you collected a particular outcome (e.g., the number of heads when flipping a coin 10 times). You also learned how to create a case table to collect the results from each trial into, and how to plot those results.

In this activity, you are going to be introduced to the `Collect` function in TinkerPlots™. This will automate the collecting of trial results in a simulation. It will also make carrying out several trials easier.

MODELING COIN FLIPS

In TinkerPlots™, set up a model to simulate tossing a single coin 10 times. (If you have forgotten how to do this, refer back to the the course activity *Modeling Random Behavior—Coin Flips*.)

- After you have set up the model, click the `Run` button.

AUTOMATING THE COLLECTION OF TRIAL RESULTS

Rather than having you record the number of heads that occurred in the 10 flips, we will automate this using TinkerPlots™. The general idea for having TinkerPlots™ record and collect the trial results is:
- Plot the outcomes from the trial.
- Collect the result.

PLOTTING THE OUTCOMES FROM A TRIAL

- Highlight the attribute `Coin` in the case table.
- Drag a new `Plot` from the object toolbar into your blank document.
- You should now have a plot with the trial's 10 outcomes color coded by whether each was *Heads* or *Tails*.
- Drag one of the circle icons in the plot until the outcomes are separated into two groups, *Heads* and *Tails*.
- Stack the outcomes using the `Vertical Stack` tool.

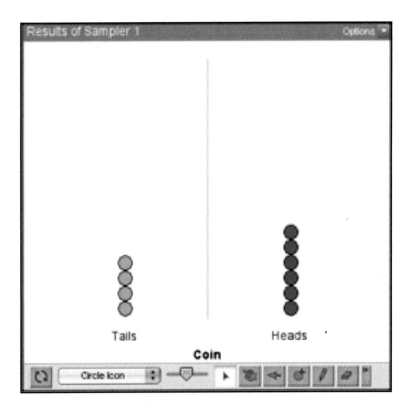

TinkerPlots™ Plot Showing the Stacked, Separated Trial Outcomes after Running the Model for Flipping a Single Coin 10 Times

COMPUTING THE TRIAL RESULT IN THE PLOT

In simulation experiments, after a trial is carried out, recall that some result from the trial is typically recorded. Remember that a result is simply a numerical summarization of the outcomes in a particular trial.

For example, in our coin flipping simulation, the result you are interested in is the *number of heads* that occurred in the 10 flips of the coin. The number of heads is a summary measure of the trial outcomes. Consider the figure below:

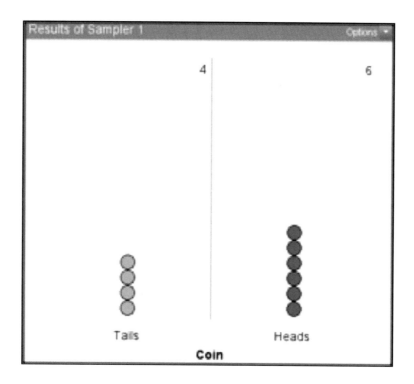

TinkerPlots™ Plot (with Counts) Showing the Stacked, Separated Trial Outcomes after Running the Model for Flipping a Single Coin 10 Times

The number of heads in this trial is six. Of course the exact same trial could have been summarized using a different measure. For example, the count of tails (4); the percentage of heads (60%); or the percentage of tails (40%) could have been used instead. The choice to summarize the outcomes using the number of heads, however, allows you to answer the questions posed in the first part of the activity.

The upper plot toolbar (see below) offers several built-in options for summarizing the plotted outcomes of a trial.

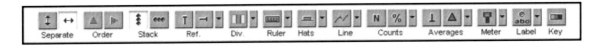

The Upper Plot Toolbar

The two Counts buttons—Count (N) and Count (%)—can be used to summarize the number and percentage of cases in a plot.

- Use Count (N) to have TinkerPlots™ count the number of heads and tails that occurred in the trial (see instructions below).

> **Summarizing the Outcomes in a Trial**
> - Highlight the plot of the trial outcomes.
> - Click on the Count (N) button in the upper plot toolbar.
>
>

Note that Count (N) and Count (%) will count the number of cases within each section of a plot. If there are not multiple sections (no bin lines), the number of total cases in the plot will be displayed.

COLLECTING THE RESULTS FROM MANY TRIALS

You can also use TinkerPlots™ to automatically collect the summarized result from the trial into a case table.

- Use TinkerPlots™ to automatically collect the result from your simulated trial into a case table (see instructions on next page).

> **Collecting the Results from a Trial**
> - Right-click the summary result in your plot.
> - Select `Collect Statistic`.

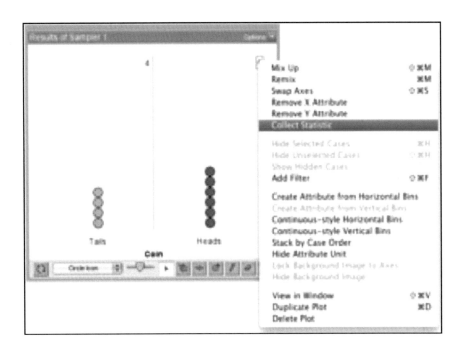

After Right-Clicking on the Summary Result in a TinkerPlots™ plot, Choose `Collect Statistic` from the Menu.

It is important that you right-click on the *actual value of the result* in the plot you want TinkerPlots™ to collect the value. For example, in the plot displayed above, you would right-click on the value *6*.

The result is then collected in a new case table. This case table, which is called *History of Results*, has a single row with the collected result, in this case six, displayed in a new attribute. The window next to the `Collect` button indicates the number of results that were collected, in this case one result was collected. This value can be changed to add the results of additional trials into the case table. In this case, the result collected from each trial is stored in a row of the *History of Results* case table.

- Change the value in the *History of Results* case table to *99* to add the results from an additional 99 trials of the simulation (see figure below).
- Click the "Collect" button.

Change the value to 99 in the History of Results table.

Use the data you collected from the 100 trials of the simulation to answer each of the following questions.

1. Record the result from the 87th trial.

2. Plot the results from your 100 simulated trials. (If you have forgotten how to do this, refer back to the instructions in the previous activity.)

3. Which result(s) for the number of heads occurred most frequently?

4. What percentage of the 100 trials had a result of *five heads*?

5. What percentage of the 100 trials had a result of *ten heads*?

6. Which result, *two heads* or *eight heads*, occurred more frequently?

MODELING DICE ROLLS

Now you will use TinkerPlots™ to simulate rolling a single die 10 times. You can again use the data you generate in the simulation to check your initial intuitions about the *number of threes* that would occur in 10 rolls.

- Open a new document in TinkerPlots™.
- Set up the model of tossing a single die 10 times.
- Carry out a single trial of the simulation.
- Plot the 10 outcomes from the simulated trial.
- Stack and separate the cases into groups.
- Use Count (N) to summarize the number of cases in each group.
- Collect the *number of threes* that occurred into a *History of Results* case table.
- Carry out an additional 99 trials.
- Plot the results from your 100 simulated trials.

Use the plot of the results from your 100 simulated trials to answer each of the following questions.

7. Sketch a plot of the results. Be sure to label the axis.

8. Which result occurred most frequently?

9. What percentage of the 100 trials had a result of *five threes*?

10. Which result, *two threes* or *eight threes*, occurred more often?

11. What percentage of the 100 trials had a result of *ten threes*?

EXTENSIONS

12. Consider the coin flipping simulation experiment that you ran. Were you able to predict the result for any one trial? Why or why not?

13. What are you able to predict in the coin flipping simulation experiment?

14. Consider the die rolling simulation experiment that you ran. Were you able to predict the result for any one trial? Why or why not?

15. What are you able to predict in the die rolling simulation experiment?

16. Describe how this is similar to what you learned in the *iPod Shuffle* course activity.

INTRODUCTION TO STATISTICAL HYPOTHESIS TESTING

In the course activities and homework assignments you have completed to this point, you have been exploring events, phenomena, and processes by using a chance model to simulate the particular outcomes or results that are generated from a particular model. This is the same kind of process that researchers and statisticians engage in when they test statistical hypotheses. Statistical hypothesis testing is a method that can be used to evaluate the likelihood of a specified model given an observed set of data. To illustrate the ideas of statistical hypothesis testing, consider how you might go about testing a coin for "fairness".

You might have suggested something along the lines of "flip the coin many times and keep track of the number of heads and tails". Suppose you tossed the coin 100 times, which resulted in 53 heads and 47 tails. Would you say the coin is "unfair"? What if you had obtained 65 heads and 35 tails instead? Now would you say the coin is "unfair"? How about if you had gotten 84 heads and only 16 tails?

The first result of 53 heads and 47 tails probably did not seem that far fetched to you, and you probably would feel at ease saying that the coin that produced such a result is most likely "fair". On the other hand, the results of 65 heads and 35 tails—and especially 84 heads and 16 tails—likely made you feel uncomfortable about declaring the coin "fair". Why is this? It is because you had a mental model of the distribution of heads and tails that you used to evaluate the observed results.

Most people, when evaluating whether a set of coin flips comes from a "fair" coin, evaluate the observed results using a model that produces a uniform distribution of the outcomes (e.g., a 50:50 split between heads and tails).

If the observed results are close to what is predicted under the assumed "fair" coin model, the model is not disputed. For example, the result of 53 heads from 100 flips is very close to the 50:50 split of heads and tails, and it is probably safe to say that a "fair" coin could have produced the set of flips in question. The other two sets of results, however, did not conform to what was expected if the 50:50 model generated 100 outcomes. This might lead you to reject the model of a "fair" coin producing these outcomes.

MODELS AND HYPOTHESES

The key to drawing an inference about whether or not the set of observed coin flips might have been generated from flipping a "fair" coin was the choice of a particular model that we used to evaluate those results against—in this case the 50:50 model. The process had us (1) select a particular model to use in the evaluation, (2) use the selected model to predict/generate the outcomes or results that would be expected under this model, and (3) evaluate the observed results within the set of expected outcomes produced from the selected model. If the observed results do not conform to what is expected, they act as evidence to dispute the model that was initially selected. We will use a simplification of this process throughout the remainder of the course:

- Model
- Simulate
- Evaluate

This may sound like a straight-forward process, but in practice it can actually be quite complex—especially as you are reading research articles and trying to interpret the findings. First off, the model that is selected is often not provided, nor described, explicitly within most research articles. It is often left to the reader to figure out what the assumed model was. At first, this may be quite difficult, but like most tasks, as you gain experience in this course and as you read more research, you find that there are a common set of models that are typically used by researchers.

The selection of a particular model is typically related to the research question being asked. Often researchers explicitly state **hypotheses** about their research questions. Hypotheses are simply statements of possible explanations for an observed set of data. For example, one possible explanation for the observed set of coin flips is that *the coin used to generate the set of observed coin flips is a "fair" coin that produces a uniform distribution of heads and tails*.

One complication that you may encounter is that many statisticians and researchers write their hypotheses mathematically. The advantage to writing a hypothesis mathematically is that it explicitly states the model. Consider the stated hypothesis that *the coin used to generate the set of observed coin flips is a "fair" coin that produces a uniform distribution of heads and tails*. Recall that producing a uniform distribution of heads and tails means that heads and tails are equally likely under this model (i.e., a 50:50 split). We could express this hypothesis more mathematically as: The proportion of heads that we would expect under this model is 0.5. Symbolically, we would express this hypothesis as:

$$H_0: \pi = 0.5$$

The symbol H_0 is common and indicates a hypothesis about a model. Here, π is the greek letter pi and means "proportion". (Typically in symbolic notation for hypotheses, π is not the mathematical constant of 3.14.) Greek letters are used when the parameters of a model are being described. In this hypothesis, we are establishing that the model we are evaluating generates heads 50% of the time.

Notice that by expressing the hypothesis mathematically, we have fully described the model we are evaluating within the hypothesis. Although the more qualitative hypothesis that *the coin used to generate the set of observed coin flips is a "fair" coin that produces a uniform distribution of heads and tails* may seem more understandable, the symbolic notation acts as a shorthand to quickly state the same hypothesis.

If this all seems like gibberish to you right now, don't worry about it. You can always write hypotheses descriptively without resorting to the symbolic notation. We only include this here to point out that all three manners of stating the hypotheses are equivalent and you will likely see different expressions of hypotheses as you read research articles or take other courses.

CONNECTIONS

In the upcoming course activities, you will explore this process of testing statistical hypotheses. You will be introduced to several common models that are assumed by researchers and statisticians. You will also use TinkerPlots™ to generate simulated data that would be expected under these models. Many of these models are directly related to the chance models that you have explored in the course to this point. For example, you should already be able to use TinkerPlots™ to produce results that would be expected from 100 flips of a "fair" coin. Aside from learning about some of the more common models used in research, you will also learn how to quantify the evidence an observed result provides against the assumed model, and how to use that evidence to evaluate a particular hypothesis and in turn answer a research question.

Greek vs. Roman Letters

Greek letters are used when the parameters of a model are being described. In contrast, roman letters are used to describe observed results. For example, go back to the situation in which the observed data consisted of 53 heads and 47 tails from 100 flips of a coin.

Here we would say $p = 0.53$. The model we are evaluating produces 50% heads, so $\pi = 0.50$.

Rather than use the roman letter, some statisticians prefer to put a "hat" on the greek letter to refer to the observed result. For example,

$\hat{\pi} = 0.53$

In this course we are not as concerned about which notation you use. In fact, it might be less confusing if you just write, *the observed result is 0.53.*

HELPER OR HINDERER READINESS

During the next class activity, you will examine results from an experiment published in *Nature* on the social evaluation of infants. To prepare for this, we would like you to read the abstract of the article so that you have an understanding of the study that was carried out.

- Read the abstract of the article *Social Evaluation by Preverbal Infants*. The abstract is available at http://www.nature.com/nature/journal/v450/n7169/full/nature06288.html

- Watch the videos that were shown to the infants in the experiment at http://www.yale.edu/infantlab/socialevaluation/Helper-Hinderer.html

Helping and hindering habituation events. On each trial, the climber (red circle) attempts to climb the hill twice, each time falling back to the bottom of the hill. On the third attempt, the climber is either bumped up the hill by the helper (left panel) or bumped down the hill by the hinderer (right panel).

COURSE ACTIVITY: HELPER OR HINDERER

Most college students recognize the difference between naughty and nice, right? What about children less than a year old—do they recognize the difference and show a preference for nice over naughty? In a study reported in the November 2007 issue of *Nature*[10], researchers investigated whether infants take into account an individual's actions towards others in evaluating that individual as appealing or aversive, perhaps laying for the foundation for social interaction. In one component of the study, 10-month-old infants were shown a "climber" character (a piece of wood with "google" eyes glued onto it) that could not make it up a hill in two tries. Then they were alternately shown two scenarios for the climber's next try, one where the climber was pushed to the top of the hill by another character (*helper*) and one where the climber was pushed back down the hill by another character (*hinderer*). The infant was alternately shown these two scenarios several times. Then the child was presented with both pieces of wood (the helper and the hinderer) and asked to pick one to play with. The researchers found that 14 of the 16 infants chose the helper over the hinderer.

---

[10] J. K. Hamlin, K. Wynn, & P. Bloom. (2007). Social evaluation by preverbal infants. *Nature, 450,* 557–559.

In this activity, you will be exploring the following research question:

> Are infants able to notice and react to helpful or hindering behavior observed in others?

DISCUSS THE FOLLOWING QUESTIONS

1. What proportion of these infants chose the helper toy?

2. What does that suggest about the answer to the research question? Explain.

Suppose for the moment that the researchers' conjecture is wrong, and infants *do not* really show any preference for either type of toy. In other words, infants just randomly pick one toy or the other, without any regard for whether it was the helper toy or the hinderer. This, remember, is the model based on random chance—the "just-by-chance" model.

3. If this is really the case (that infants show no preference between the helper and hinderer), is it possible that 14 out of 16 infants could have chosen the helper toy just by chance?

4. Do you think that the observed result (14 of 16 choosing the helper) is a **likely result** if infants had no real preference, or would it be an unlikely result? How strong do you believe the evidence is against the "just-by-chance" model if 14 of 16 infants were found to choose the helper?

THE "JUST-BY-CHANCE" (NO PREFERENCE) MODEL

Consider the argument that that infants have no preference for either the helper or the hinderer. If that argument is true, then you would expect that any infants selecting the helper is only because of random chance—not due to any underlying psychological tendency of infants to prefer helpers to hinderers.

The good news is that this "just-by-chance" process (i.e., random chance) can be modeled using the same chance devices that you have been using in the course thus far. Under the assumption of "just-by-chance", the process of infants selecting toys can be modeled by randomly selecting either a helper or a hinderer. After the random selection, it can be determined how many infants "chose" the helper. This process can be repeated a large number of times to simulate the percentage of infants selecting the helper under this "no preference" model.

SUMMARY OF THE SIMULATION PROCESS

The key to answering the research question in this activity is to determine the likelihood of the observed result (14 of 16 infants choosing the helper) under the assumption that infants have no preference for either the helper or the hinderer. The "no preference" model is again the "just-by-chance" model—infants randomly select either the helper or hinderer.

To determine this likelihood, you will model the process of 16 hypothetical infants making their selections using random chance. Then, you can count how many of these "infants" choose the helper toy. This process can be repeated many times to obtain a distribution of results that would be expected under the "no preference" or "just-by-chance" model.

The observed result of 14 of 16 infants choosing the helper can then be evaluated in light of this distribution to determine how likely it would be to obtain such a result (or a more extreme result) under the assumption of random chance. As such, the observed result can provide evidence to help answer the research question.

5. Draw a picture of the sampler (model) that you will use to generate outcomes from the "just-by-chance" model. In the picture, be sure to (1) indicate the type of sampling device used (mixer, spinner, etc.); (2) label all the elements in your sampling device; (3) label the probability associated with each element; and (4) indicate the Repeat and Draw values you will use.

SIMULATING THE DATA

In this study, a trial represents each of the 16 infants choosing a toy. The trial ends when 16 toys have been chosen randomly.

- Carry out a single trial of the simulation in TinkerPlots™.
- Plot the outcomes from the trial.

6. Sketch a plot of the outcomes from this trial. Add all labels and summary measures (counts, percentage, etc.) to your plot.

7. Based on your sketch, what, specifically, is the result that you will be **collecting**?

- Collect the appropriate summary measure.
- Carry out 499 more trials (500 trials total) of the simulation in TinkerPlots™.

EVALUATING THE HYPOTHESIZED MODEL

- Plot the results from the simulation.

8. Sketch a plot of the results below.

9. Is the observed result from the original experiment likely or unlikely under the hypothesized model? Explain.

10. Does the observed result support the hypothesized "just-by-chance" model? Explain.

QUANTIFYING THE LEVEL OF SUPPORT FOR A MODEL BASED ON THE OBSERVED RESULT

Consider what it means for the observed result to suggest support for the hypothesized model. Recall the example of the hypothesized model from the Introduction to Statistical Hypothesis Testing reading: You have a sequence of observed coin flips and want to know if the observed sequence might have been generated from a "fair" coin (H_0: π = 0.5). Suppose you flip your "unknown" coin 50 times and observe 27 heads. How much support does this result provide for your hypothesized model of a "fair" coin?

The first step is to create a model of a fair coin and generate many simulated trials of flipping a "fair" coin 50 times and observing the number of heads. Then, you can plot your simulated trials. Obviously, if many of the simulated trials are exactly the same as the observed result, that would provide quite strong support for the hypothesized model.

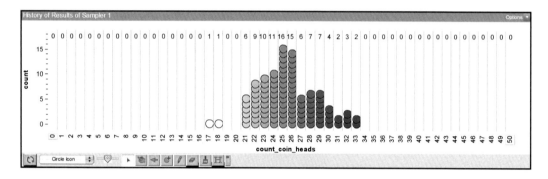

Simulation results for 100 simulated trials of flipping a "fair" coin 50 times.

11. What proportion of the 100 simulated trials, above, produced a result of exactly 27 heads? Does this mean that 27 heads is an unlikely result under the hypothesized model of a "fair" coin? Explain.

We expect about 25 heads if the coin is truly "fair", so it is not only relevant if the simulated trials produce exactly 27 heads. If many simulated trials also produce results of 28, 29, or higher this actually provides even stronger support for the hypothesized model since it implies that even higher numbers of heads than 27 are likely under the hypothesized model.

Therefore when we want to evaluate our model, we should consider an evaluation region—the range of possible results that we can use to evaluate whether the hypothesized model is supported by the observed results. If there are many simulated results in the evaluation region, then the hypothesized model is well-supported by the observed result. If there are few, we may conclude that our hypothesized model is not plausible. For the above example, the evaluation region is 27 heads or more since simulated results that fall into this range provide support for the hypothesized model.

The level of support for a model is quantified by answering the question: **What proportion of the simulated results fall into the evaluation region?** The figure below displays the evaluation region for our example of flipping an "unknown" coin 50 times and observing 27 heads. The percentage of evaluation region constitutes the proportion of simulated results where we obtained 27 heads or more. This indicates the level of support for the hypothesized model given the observed result of 27 heads.

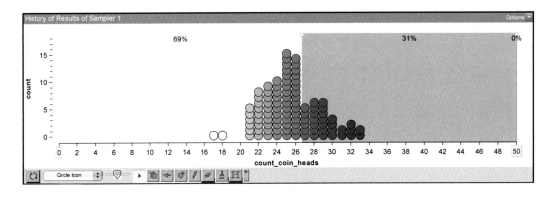

Evaluation region for our example of flipping an "unknown" coin 50 times, observing 27 heads.

The results of 28 or more heads are also included since if someone considers simulated results that are the same as the observed result of 27 as evidence to support the hypothesized model, then results of 28 must also be evidence to support the hypothesized model, as must ALL results more than 27 or 28.

The level of support is usually expressed as a proportion (not a percentage). In this example, the level of support based on the observed result of 27 is 0.31. This tells us that the observed result of 27 is in the most extreme 31 percent of the simulated results. This is a fairly likely result given the hypothesized model and therefore provides a moderate to high level of support for the hypothesized model (since it could happen nearly than 1 in 3 times if we ran the experiment with a "fair" coin). We would conclude that we do not have strong evidence that our coin is "unfair" solely based on the fact that we observed 27 heads.

12. Based on the evaluation region, quantify the level of support for the hypothesized model based on the observed result of 14 out of 16 infants choosing the helper toy based on the hypothesized model of no difference (see instructions below).

Using the Divider Tool to Quantify the Level of Support for a Model Based on an Observed Result
- Highlight the plot of the results.
- Click on the Divider tool in the upper plot toolbar.

- Move the divider by dragging the end-lines so that the grey area covers the evaluation region. This region extends from the observed result to the most unlikely result under the hypothesized model. The proportion of simulated results in this range provides a measure of how well the hypothesized model can account for the observed result. In this example, the grey area would extend from 14 to 16.
- Click the Counts (%) button.

13. Given the level of support you just computed, does this suggest that the hypothesized model is supported or not? Explain.

14. What does this suggest about the research question? Is it likely that infants are making their selections based only on random chance? Explain.

Support of the Model

It is important to remember that although we can rule out a model based on low evidence for support, evidence that yields high support for the model does not mean the model is correct! A particular observed result might support several models…and we only examined one particular model, the "just-by-chance" model. (Later in the course we will learn how we can determine which models a particular observed result will support.)

We can never claim a model is correct based only on the results of a statistical test. Because of this we have to be very careful with the language we use when we describe our statistical results. (When there is a high degree of evidence found, many statisticians say "we **cannot reject** the model".)

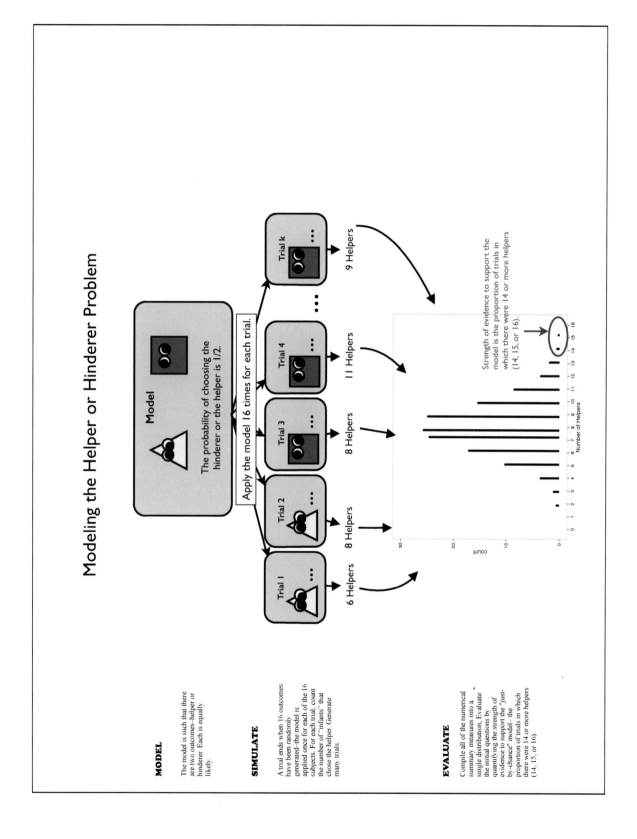

MISSION IMPROBABLE: A CONCISE AND PRECISE DEFINITION OF P-VALUE

During the last course activity, *Helper or Hinderer,* you learned how to quantify the level of support for the hypothesized model. This quantification is called a *p*-value. The article below gives more information about *p*-values and a rather arbitrary criteria (if $p < 0.05$) that some researchers use as *non-support* for a hypothesized model.

- Read the article *Mission Improbable: A Concise and Precise Definition of p-Value*
  http://news.sciencemag.org/2009/10/mission-improbable-concise-and-precise-definition-p-value

COURSE ACTIVITY: MONDAY BREAKUPS

Facebook is a social networking website. One piece of data that members of Facebook often report is their relationship status: single, in a relationship, married, it's complicated, etc.

With the help of Lee Byron of Facebook, David McCandless—a London-based author, writer, and designer—examined changes in peoples' relationship status, in particular, breakups. A plot of the results showed that there were repeated peaks on Mondays, a day that seems to be of higher risk for reported breakups.

Consider a random sample of 50 breakups reported on Facebook within the last year. Of these sampled breakups, 13 occurred on a Monday.

In this activity, you will be exploring the following research question:

> Is the percentage of breakups reported on Mondays higher than we would expect from random chance?

DISCUSS THE FOLLOWING QUESTIONS

1. What percentage of breakups were reported on a Monday?

2. What does that suggest about the answer to the research question? Explain

Suppose for the moment that the researchers' conjecture is wrong, and breakups *are not* reported on Monday more than any other day. In other words, breakups are reported randomly throughout the week. This, again, is the model based on random chance—the "just-by-chance" model.

3. Based on the "just-by-chance" model, what percentage of breakups would we expect to be reported on Mondays? Using this value, write the null hypothesis for the "just-by-chance" model.

> **Null Hypothesis**
>
> The null hypothesis is a statement that describes the "just-by-chance" model.

4. Draw a picture of the sampler (model) that you will use to generate outcomes from the "just-by-chance" model. In the picture, be sure to (1) indicate the type of sampling device used (mixer, spinner, etc.); (2) label all the elements in your sampling device; (3) label the probability associated with each element; and (4) indicate the Repeat and Draw values you will use.

- Set up the model/sampler in TinkerPlots™.

SIMULATING THE DATA

- Carry out a single trial of the simulation in TinkerPlots™.
- Plot the outcomes from the trial.

5. Sketch a plot of the outcomes from this trial. Add all labels and summary measures (counts, percentage, etc.) to your plot.

6. Based on your sketch, what, specifically, is the result that you will be collecting?

- Collect the appropriate summary measure from the plot of your first trial.
- Carry out 499 more trials (500 trials total) of the simulation in TinkerPlots™.

EVALUATING THE HYPOTHESIZED MODEL

- Plot the results from the simulation.

7. Sketch a plot of the results below.

8. Quantify the level of support for the "just-by-chance" model based on the observed result (i.e., report the p-value).

9. Given the level of support you just computed, does the observed result support the hypothesized model or not? Explain.

10. Use the evidence from the simulation (e.g., the level of evidence) to answer the research question.

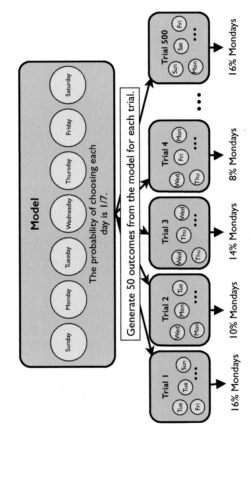

COURSE ACTIVITY: RACIAL DISPARITIES IN POLICE STOPS

In the 1990s, the U.S. Justice Department and other groups studied possible abuse by Philadelphia police officers in their treatment of minorities. One study conducted by the American Civil Liberties Union analyzed whether black drivers were more likely than others in the population to be targeted by police for traffic stops.

Researchers studied the results of 262 police car stops during one week in 1997. Of those, 207 of the drivers were black. At that time Philadelphia's population was 42.2% black.

In this activity, you will be exploring the following research question:

> Does the percentage of black drivers being stopped provide evidence of possible racial disparities (i.e., higher than what we would expect because of chance variation)?

DISCUSS THE FOLLOWING QUESTIONS

1. What percentage of the drivers that were stopped were black?

   79 percent were black

2. What does that suggest about the answer to the research question? Explain

   That there is some racial disparity in the police stops

Null ↓

Suppose for the moment that there are *no* racial disparities in police stops. In other words, the percentage of blacks stopped by the police should reflect the percentage of blacks in the population.

3. What percentage of the drivers that were stopped would you expect to be black, given the population?

   42.2 percent

In both the *Helper or Hinderer* course activity and the *Monday Breakups* course activity, all of the elements in the "just-by-chance" model had the same probability. When the probabilities of each element in the model are exactly the same, we say the "just-by-chance" model is a **uniform probability model**.

The "just-by-chance" model does not have to be a uniform probability model. The elements can have differing probabilities. As long as the elements being selected are still random, this is still a "just-by-chance" model.

4. Using the value from the previous question, write the hypothesis for the "just-by-chance" model.

In the long run, 42.2 percent of drivers stopped by Phil. police would be black.

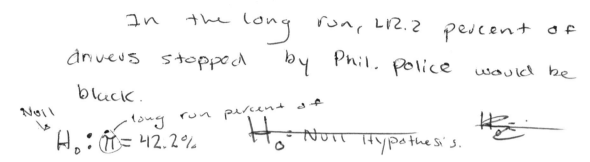

$H_0: \pi = 42.2\%$

5. Draw a picture of the sampler (model) that you will use to generate outcomes from the "just-by-chance" model. In the picture, be sure to (1) indicate the type of sampling device used (mixer, spinner, etc.); (2) label all the elements in your sampling device; (3) label the probability associated with each element; and (4) indicate the Repeat and Draw values you will use.

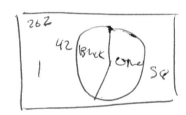

- Set up the model/sampler in TinkerPlots™.

SIMULATING THE DATA

- Carry out a single trial of the simulation in TinkerPlots™.
- Plot the outcomes from the trial.

6. Sketch a plot of the outcomes from this trial. Add all labels and summary measures (counts, percentages, etc.) to your plot.

Other Black

7. Based on your sketch, what, specifically, is the result that you will be collecting?

- Collect the appropriate summary measure from the plot of your first trial.
- Carry out 499 more trials (500 trials total) of the simulation in TinkerPlots™.

EVALUATING THE HYPOTHESIZED MODEL

- Plot the results from the simulation.

8. Sketch a plot of the results below.

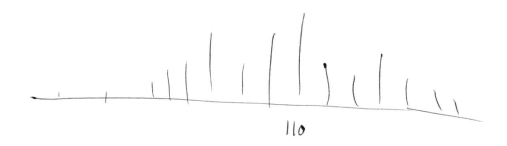

110

9. Quantify the level of support for the "just-by-chance" model based on the observed result (i.e., report the *p*-value).

$$\frac{0}{500} \qquad 0 \qquad \frac{\text{\# of results as extreme or more extreme}}{\text{total results}}$$

10. Given the level of support you just computed, does the observed result support the hypothesized model or not? Explain.

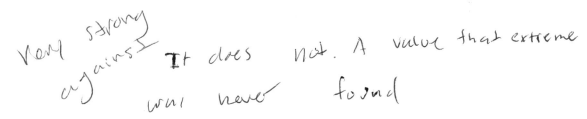

Very strong against. It does not. A value that extreme was never found.

11. Use the evidence from the simulation (e.g., the level of evidence) to answer the research question.

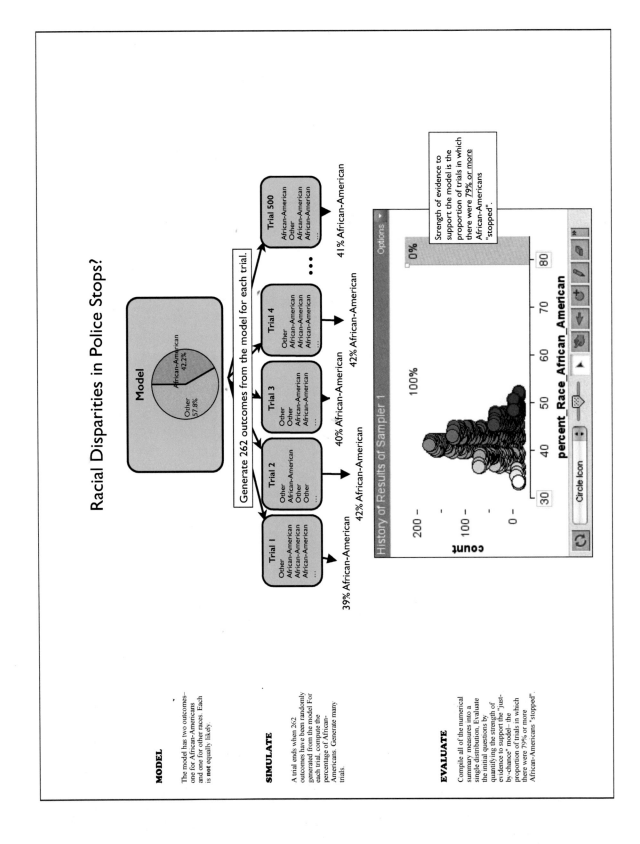

LEARNING GOALS: UNIT 1

The activities, homework, and reading that you have completed for the first part of this course have introduced you to several fundamental ideas in the discipline of statistics including randomness, modeling, and simulation. You have also been introduced to the key elements that are used by researchers and scientists for examining and testing conjectures and hypotheses. In addition, you have learned how to use TinkerPlots™ to model probabilistic events and generate simulated data under a variety of conditions.

These ideas and skills are crucial for your success in the remainder of this course, as they will be extended and built on in the upcoming material. With that in mind, presented here is a list of some key concepts and skills that you should have a good handle on.

LITERACY/UNDERSTANDING (TERMS AND CONCEPTS)

You should understand that:

- Human intuitions about randomness/probability may be faulty.
- Randomness/probability cannot be out-guessed in the short term but patterns can be observed over the long term.
- Simulation can be used to investigate probabilistic outcomes and model chance events.
- Simulation can be used to determine whether a particular result is likely to have happened "just by chance.
- Different chance models lead to different simulation results
- There are predictable patterns/characteristics of simulation results based on repeatedly sampling/generating random data (e.g., a bell shape from a graph of sample statistics).
- An observed result can be used as evidence to support/not support a hypothesized model.

MODEL-SIMULATE-EVALUATE FRAMEWORK

You should be able to:

- Translate real life phenomena into a model to be used in the simulation process.
- Use TinkerPlots™ to generate random outcomes to simulate experiments/real-life phenomena under the "just-by-chance" model.
- Use an observed result as evidence to evaluate the support for a hypothesized model.

TINKERPLOTS™ SKILLS

You should also be able to do all of the following using TinkerPlots™:

- Create a new case table and enter data into the case table.
- Create a new sampler and use different devices to generate random outcomes (e.g., `Spinner`, `Mixer`, `Counter`).
- Plot values from an attribute in a case table and organize (by separating) the the plotted values.
- Numerically summarize the randomly generated outcomes from the trial (e.g., `Count (N)`, `Count (%)`).
- Collect the numerical summary measure from many trials.
- Use the `Divider` tool to compute the *p*-value.

In the next activity, *Unit 1 Wrap-Up & Review*, you will have a chance to evaluate whether or not you have mastered these ideas through a variety of practice and extension problems. As a pre-cursor to this activity, you may want to review the readings and course activities from Unit 1.

COURSE ACTIVITY: UNIT 1 WRAP-UP & REVIEW

TERMINOLOGY FOR UNIT 1

At this point, you should be familiar with the following terms. Write down what each term represents as well as any notes that may help you remember.

1. Model

2. Trial

3. Observed result

4. Random chance ("just-by-chance") model

5. Null hypothesis

6. Level of support/p-value

MODELING REAL-WORLD PHENOMENA

7. Suppose that you receive ten text messages per day and one-third (1/3) of all text messages come from your mother. Describe how you could use TinkerPlots™ to simulate the number of text messages that you receive from your mother each day.

8. Carry out your described simulation using TinkerPlots™ to simulate receiving 30 days worth of texts. Sketch the distribution for the number of texts received from your mother.

P-VALUE

Suppose that the observed result from the *Helper or Hinderer* research study had been 10 of the 16 infants choosing the helper toy (rather than 14 of 16).

9. Explain why you can obtain the *p*-value for this new observed result using the results from the simulation analysis that you already conducted in the *Helper or Hinderer* course activity.

10. Based on your simulation analysis from the *Helper or Hinderer* activity, quantify the amount of support for the "just-by-chance" model based on this new observed result (i.e., report the *p*-value).

11. What conclusion would you draw about the "just-by-chance" model? Explain your reasoning process behind this conclusion.

12. If the new observed result had been 13 of 16 choosing the helper toy, quantify the the amount of support for the "just-by-chance" model and draw a conclusion about the model.

Suppose that the study had involved only eight infants rather than 16, and that seven of the eight infants had chosen the helper toy.

13. Explain how you would modify the simulation analysis for this new study.

14. Would you expect the observed result from this new study to constitute more, less, or the same amount of evidence that infants really prefer the helper toy, as compared to the original study result (14 out of 16 prefer the helper toy)? Explain your answer.

15. Carry out the simulation analysis and comment on whether your expectation was correct based on the results.

FURTHER PRACTICE

A recent study investigated whether couples had a leaning preference when kissing. The researcher found that eight of 12 couples leaned their heads to the right when kissing. To examine whether or not couples had a right-leaning preference, the researchers used TinkerPlots™ to simulate from the "no preference" model. They simulated data for 500 trials. In each trial, the number of "couples" who lean their heads to the right was collected. The plot of these results is shown below.

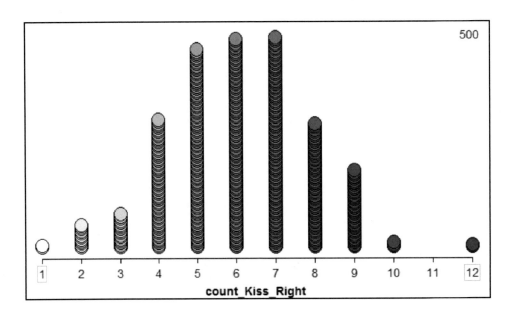

Simulation results based on 500 trials.

16. Describe how to set up a model in TinkerPlots™ to obtain these results.

17. Use the plot of the simulated results and the observed result to quantify the level of support for the "no preference" model (i.e., report the *p*-value).

18. Based on your quantification, does the evidence support the hypothesis that couples have no leaning preference when kissing? Explain.

In 2008, the British mobile company Three reported that 76% of 18–24 year olds were guilty of communifaking—the practice of 'simulating social interaction via telephone in order to avoid looking like a Billy-no-mates' (i.e. pretending to be on the phone when you're not). To examine whether or not University of Minnesota students engage in less communifaking than the general public, an informal poll was conducted in a class of 22 University of Minnesota students. This poll found that only 15 of the students had engaged in communifaking.

Conduct a simulation analysis to determine whether or not there is statistical evidence to suggest that University of Minnesota students engage in less communifaking than the general public.

19. Sketch a plot of the simulation results and quantify the level of support for the "just-by-chance" model based on the observed result (i.e., report the p-value). Then, use the evidence from the simulation (e.g., the level of evidence) to answer the research question.

UNIT II: COMPARING GROUPS

The nature of doing science, be it natural or social, inevitably calls for comparison. Statistical methods are at the heart of such comparison, for they not only help us gain understanding of the world around us but often define how our research is to be carried out.

—T. F. Liao (2002)

Drawing inferences and conclusions about the differences among groups is an almost daily occurrence in the lives of most people. In any given hour of any given day, television, radio and social media abound with comparisons. For example, data scientists at *OKCupid*, an online dating site, examined whether frequent tweeters (users of Twitter) have shorter real-life relationships than others[11].

Group comparisons are at the heart of many other interesting questions addressed by psychologists, physicians, scientists, teachers, and engineers. Aside from questions of differences, group comparisons address questions about efficacy, such as: "Is a particular curriculum effective in improving students' achievement?" or "Is having a Facebook page for a business effective in improving sales?"

---

[11] The website OKTrends includes an answer to this question, as well as many others.

They also address questions about magnitude such as: "How much lower are attendance rates for a particular population of students?" or "How much more likely is it that an iPhone user will live in the city (rather than the suburbs or the country) than an Android user?[12]"

COMPARISONS IN THE MEDIA

The media often proffer group comparisons. Consider the infographic[13] displayed in the figure below. It compares survey data from cat and dog people.

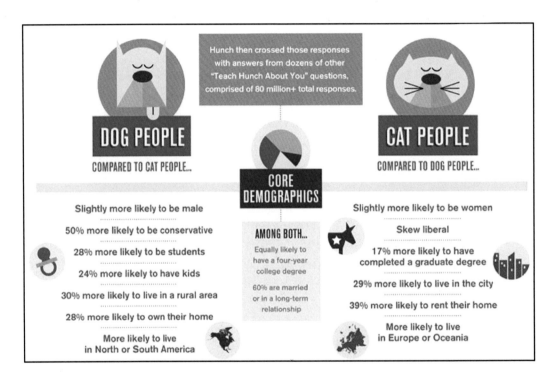

Hunch infographic comparing dog and cat people.

The comparison between cat and dog people is primarily conveyed via text. The graphical parts of the infographic are mostly aesthetic.

The figure presented on the next page shows a screenshot from a Facebook application called *Facebook Questions*. The application allows any Facebook

---

[12] This question was asked in a survey by *Hunch*, a personal recommendation service. They created an infographic that highlighted several comparisons of the average demographic characteristics between the two users.

[13] The entire infographic can be seen at http://blog.hunch.com/?p=50354.

user who has installed it to poll other Facebook users. It displays the poll results as a bar graph. Hovering over a particular bar will display the frequency or count of users voting for a particular response.

In contrast to the comparison in the *Hunch* infographic, the comparison here is conveyed through the graph rather than the text. The length of each bar provides an indication of the frequency of voters who responded to each of the poll options relative to the other options. For example, the bar labeled 'Eric' in the *True Blood* poll is roughly five times longer than the bar labeled 'Goderic'. This corresponds to the five-fold difference in the frequency of votes (3,571 votes for Goderic and 18,102 votes for Eric) between the two candidates. The bar labeled 'Bill' is roughly 1/2 the length of the bar labeled 'Goderic' and about 1/9 as long as the bar labeled 'Eric', indicative of the 1,933 votes that had been cast for Bill.

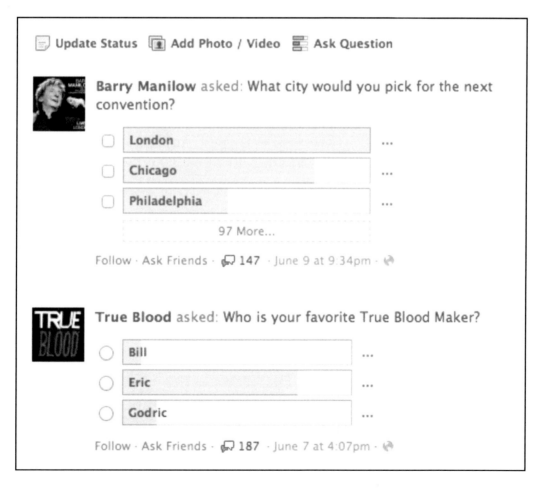

Poll questions and bar graphs displaying the voting results for two questions submitted using the *Facebook Questions* application.

Examining the two graphical displays presented, reflect on the following questions.
- Which of these two displays more clearly conveys the comparisons being made? Why?
- How would you improve on each display? Explain.

STATISTICAL COMPARISONS

Statisticians have made many contributions to the methodology used by researchers and scientists in making group comparisons. One well-known statistician, Roger Kirk, has suggested that research questions regarding group differences should address three important questions: (1) Is an observed difference real or should it be attributed to chance? (2) If the difference is real, how large is it? and (3) Is the difference large enough to be useful?

Each of these three questions can be addressed through statistical inference. Recall that statistical inference is the process through which we can deal with uncertainty. Consider the following research questions posed by the *Sunlight Foundation*: Has the complexity of speeches given in the United States congress changed over the last 10 years? Is the complexity of speeches given by Democrats different than that of speeches given by Republicans?

OPERATIONALIZATION

In thinking about answering these questions, there are several decisions that need to be made by the researcher. One major decision that would have to be made is how "complexity" is to be measured. This **operationalization** involves defining and quantifying a fuzzy construct to help make it understandable (and study-able). In general, making a comparison between numeric properties such as three apples and five apples is much easier than making comparisons of an ill-defined characteristic such as "complexity". In considering how to operationalize a construct, there are often many possibilities. For example, complexity of a written passage can be operationalized by using a readability index such as the Flesch–Kincaid score. Another way of operationalizing the passage is to count the number of words in the passage that appear in Kaplan's list of *The 100 Most Common SAT Words*[14].

---

[14] See Kaplan's entire list of *The 100 Most Common SAT Words* at http://www.washingtonpost.com/wp-adv/eduadv/kaplan/kart_ug_sat100.html.

It is important to understand the advantages and limitations in the choice of an operationalization. For example, by using the frequency of the 100 most common SAT words, it is easy to provide a quantification for each passage. It is also easy to understand what that quantification means—a passage with a score of four uses two fewer SAT words than a passage that has a score of six. This will ease interpretations during an analysis. On the other hand, this score is a rather rudimentary measure of complexity.

In comparison to the frequency of SAT words, the Flesch–Kincaid score is a less rudimentary measure of sentence complexity defined as,

$$FK = 0.39 \times (\text{average words per sentence}) + 11.8 \times (\text{average syllables per word}) - 15.59.$$

The score indicates that the text would be at the limit of comprehension for a person with the equivalent of that number of years of education[15]. In more familiar words, the score gives the reading grade level of the passage.

> Many famous passages have been scored using the Flesch–Kincaid measure such as the U.S. Constitution (17.8 grade level), the Declaration of Independence (15.1 grade level), the Gettysburg Address (11.2 grade level), Martin Luther King's "I Have a Dream" speech (9.4 grade level), and U2 frontman Bono's 2004 commencement speech at the University of Pennsylvania (5.9 grade level).

At its core, the Flesch–Kincaid score equates higher grade levels with longer words and longer sentences. It does not, however, indicate anything about the clarity or correctness of a passage of text. If these attributes are important in operationalizing passage complexity, other measures should be used in the quantification. Another strength of the Flesch–Kincaid score is that it has been adopted as the Department of Defense standard in determining the age-

---

[15] For example, a passage with a Flesch–Kincaid score of 12 indicates that a person with 12 years of education would answer 50 per cent of the questions correctly.

appropriateness of reading material. Because of this, the metric is one of the best known and frequently used metrics of readability.

The choice of operationalization is subjective to the researcher and inevitably has a great deal of impact on the generalizations that the researcher can make. Does the operationalization accurately reflect the construct that the researcher wishes to say something about? For example, say that after making comparisons of congressional speeches using the Flesch–Kincaid score you find that more current speeches have, in general, lower scores than speeches from 10 years ago. This might suggest that the complexity of today's congressional speeches is less than it was 10 years ago. Another researcher might choose a different method of operationalizing complexity and come to a different conclusion. What does the choice of operationalization reflect about the "truth" of the change (or non-change) of complexity in congressional speeches? In science, these types of questions are related to the **validity of the inferences** that can be made[16].

SUMMARIZATION

After deciding upon an operationalization, it may seem that a researcher could just collect her data and answer her research questions. However, there are still a litany of choices that the researcher needs to make before she gets to this point. For example, again consider the two research questions posed previously: Has the complexity of speeches given in the United States congress changed over the last 10 years? Is the complexity of speeches given by Democrats different than that of speeches given by Republicans?

One of the initial questions is whether the researcher should use **raw data** or **summary data**. Raw data would be data that has not been processed or manipulated in any way (i.e., the original data). In this case, it would mean using the transcripts of the speeches given in Congress. Summary data, on the other hand, would refer to data that has been manipulated or summarized in some manner. For example, rather than obtaining a transcript of the actual speeches given, the researcher could instead use data that some other party or individual has already summarized and produced such as the average word per sentence or even the Flesch–Kincaid scores themselves. Most researchers would prefer to work from the raw data and compute these measures

---

[16] Many books on research methods deal extensively with ideas of validity. The Research Methods Knowledge Base provides more information for the interested reader.

themselves, but in many cases the raw data is not available and only summary information is accessible[17].

In the research proposed, It is in fact possible for the researcher to obtain the raw data—transcripts for each of the Congressional speeches given in the last 10 years. The Congressional Record is a verbatim account of the remarks made by senators and representatives while they are on the floor of the Senate and the House of Representatives. Transcripts for each speech could be collected from this, or another, website. Again, the collection of these transcripts may seem straight-forward (just download them from *The Congressional Record*). However, just a minute or two of examining the website leads to many other questions. For example, after searching on "Amy Klobuchar", one of the Senators from Minnesota, it was revealed that the search results brought up any mention of Amy Klobuchar—not just the speeches she gave. Should we use all of these? Or just the speeches? Another decision.

It would also soon be apparent that collecting these data could be very time consuming. There were 219 search hits for Amy Klobuchar in the 112th Congress (2011--2012) alone. Given there are 541 members of Congress in any given year (435 voting Representatives, 6 non-voting Representatives, and 100 Senators), and we would need 10 years worth of data, this is somewhere on the magnitude of 1,082,000 transcripts that we would need to download[18]. (Not to mention that we would also have to compute the Flesch–Kincaid score for each one!). Because of this, many researchers collect only a subset of the data available. This is referred to as **sampling** and there are well-defined methods for choosing a sample of data to ensure that it adequately represents the entire population.

After all of the decisions about operationalization and data collection have been made, and the transcripts have been collected and quantified with a Flesch–Kincaid score we can start to think about analyzing the data to answer

---

[17] As an example, the Minnesota Department of Education is required to provide test results, revenue and expenditure data, and demographic information for each school and district in the state. Because of privacy laws, these data are released to the public at the school-level (e.g., summarized as school averages) rather than at the individual-level (i.e., the raw data used to compute these averages). See the MDE website for reports and publicly available data for all Minnesota schools and districts.

[18] Some of the downloading of documents and processing (e.g., counting syllables, words per sentence, etc.) can be automated using computer languages such as Python or Perl. This could still take several weeks to write the web scraping program to obtain the data and check the resulting output for errors.

the research questions. This leads to more questions and decisions that need to be made. How will you compare the Flesch-Kincaid scores across years? Between Democrats and Republicans? Consider the plot (on the next page) of Flesch-Kincaid scores for ten speeches given by Democrats and ten speeches given by Republicans.

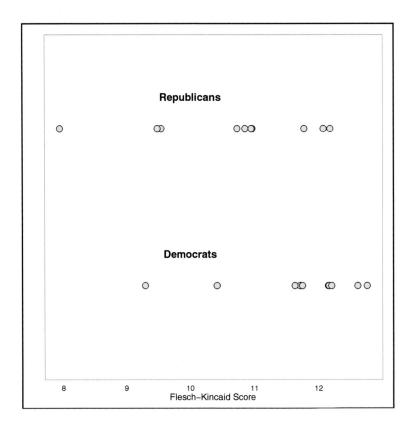

Flesch-Kincaid scores for ten speeches given by Democrats and ten speeches given by Republicans.

Using the plot, reflect on how you might answer the following questions.
- How would you compare the Flesch-Kincaid scores for the speeches given by Democrats and the speeches made by Republicans?
- If you had to summarize the Flesch-Kincaid scores for the speeches given by Democrats by using a single number to quantify all ten speeches, what would this number be? Explain.
- Are you comfortable summarizing the ten scores into a single number? Explain. If not, are there other numbers you would use or add to your original quantification? Explain.

- Using the same methodology, summarize the Flesch–Kincaid scores for the speeches given by Republicans.
- Compare the two quantifications you have just made. Summarize the differences between the Democrats and Republicans based on the two quantifications into a single number (i.e., how much bigger/smaller is the Democrats score than the Republicans?)
- What does this comparison suggest about the original research question as to whether the complexity of speeches given by Democrats are different than that of speeches given by Republicans? Explain.

TO INFER OR NOT TO INFER

It may seem that after all of this work we finally have an answer to our research questions, but it turns out that we may not. This is because we have introduced some error through the process of sampling (the data constitutes *only* 10 speeches from each party). Let's say that we had sampled 200 speeches (100 speeches given in 2002 and 100 given in 2012). Further consider that we found the average grade-level for the 100 speeches given in 2002 was 11.5 and the average grade-level for speeches given in 2012 was 10.6. This suggests that Congressional speeches currently are almost an entire grade-level lower than they were 10 years ago

But, what if we had gotten different speeches in our sample? There are many different samples of 100 speeches from 2002 and 100 speeches from 2012. Each of these different samples might give a slightly different value for the average grade-levels for those years, which in turn affects how different we claim these grade-levels are! It is well-known that different samples give different results (e.g., averages). A very big question that we will address in this unit is: **How different can the results be just because of obtaining a different sample?** By answering this question, we can address and quantify how uncertain our observed results are. For example, we could find, based on different samples, that the difference in grade-level between speeches given in 2002 and 2012 may be as high as 1.3 or as low as 0.5.

The ideas regarding the quantification of uncertainty of results because of sampling relate directly to ideas of statistical inference. Statistical inference is the ability to draw conclusions from sample data. Consider if the range of uncertainty is very large, such as the difference in grade-level between speeches given in 2002 and 2012 may be as high as 2.1 or as low as -0.2 (a negative value here indicates that the grade-level for speeches given in 2012 is actually higher than that for speeches given in 2002). This much uncertainty makes it very difficult to address how different the complexity in speeches given in 2002 and 2012, and in fact, whether there is a difference at all.

Quantifying uncertainty is a large part of statistical inference, but it is not the only thing involved in drawing conclusions from sample data. Drawing conclusions from data also implies thinking broadly about the scope of those inferences. For example, does a difference in complexity only refer to a comparison of speeches given in 2002 and 2012? Or does it apply more broadly to the complexity of all Congressional speeches ever given? Or, can we use these data to suggest that speeches in general—Congressional, presidential, commencement, etc.—have changed in complexity over time?

Another part of drawing conclusions is in considering the 'why'. Why has the complexity of Congressional speeches changed over time? Is this an indictment of Congress? the American public? or neither? Ascribing meaning to why differences occur is a part of the much broader reasons for carrying out research. Unfortunately, the attribution of the reasons or causes of a particular difference are very difficult under most circumstances.

OUTLINE OF THE UNIT

In this unit, you will learn how to quantify the uncertainty associated with the results from a sample and how that uncertainty is expressed in the reporting of research. You will also learn about how particular statistical methods influence the scope of inferences and the attribution of cause that a researcher can make. Understanding these ideas is important in evaluating any given result presented in any research that has used statistical methods.

In this unit you will be introduced to the randomization test to examine whether difference between groups observed in data are more than you would expect because of chance variation. You will also learn how to carry out a randomization test using the TinkerPlots™ software.

The ideas of the randomization test will be reinforced throughout the remainder of the unit as will its utility, as you explore data in which the outcomes are both quantitative and categorical. You will also examine data that have been collected under many different conditions, some experimental and others observational.

As you encounter data collected under these various conditions, you will learn how study design and data collection directly affects the scope of inferences that are appropriate. Throughout the unit, you will continue to add to the knowledge you accumulated from Unit 1. This includes both content knowledge and TinkerPlots™ knowledge.

REFERENCES

Kirk, R. E. (2001). Promoting good statistical practices: Some suggestions. *Educational and Psychological Measurement, 61*(2), 213–218.

Liao, T. F. (2002). *Statistical group comparison.* New York: Wiley.

AMERICA'S MOST RELIABLE AIRLINES

Based on excerpts from an article by Rebecca Ruiz in the October 1, 2008 Forbes Magazine.

According to our analysis of the nation's 10 major airlines, discount carriers actually rank first in reliability.

Southwest, the no-frills discount carrier, handily beat the competition in most of the categories we judged. JetBlue, also considered a discount airline despite its plush leather seats and individual television sets, ranked third just behind Continental Airlines. Fourth place went to AirTran, another budget carrier.

Alaska Airlines, American Airlines, and Delta Air Lines were solidly average performers. United Airlines and US Airways landed at the bottom of the list.

To judge reliability in the airline industry, particularly at a time when carriers are responding to oil prices by slashing capacity and raising prices, we looked at six different factors for 10 major airlines.

We collected five years' worth of data relating to on-time arrival, cancellations, complaints, and mishandled baggage from the Aviation Consumer Protection Division of the Department of Transportation. Delays and cancellations, the factors most likely to ruin a flier's day, were given double weight.

To better gauge the overall flying experience, we included *J.D. Power and Associates'* consumer satisfaction rankings from 2005 to 2008. These surveys reach more than 9,000 travelers annually and ask participants to rate factors like cost and fees, in-flight services, and check-in.

When all of these figures were combined, the discount airlines consistently rose to the top. For each of the years we studied, Southwest's flights were punctual more than 80% of the time; the average was 76.8%. Alaska Airlines gave the most dismal performance, with only 74.6% on-time flights.

In terms of canceled flights, Southwest reigns yet again. The carrier canceled an average of only 0.65% of its flights over the five-year period, compared with the worst airline, American, which canceled an average of 2.4%.

AirTran, another budget carrier, had the fewest reports of mishandled baggage—a contentious issue now that airlines are charging as much as $50 to check regular-sized luggage. In 2007, AirTran had about four reports of mishandled baggage per 1,000 customers. The worst-ranking airline, US Airways, had 8.5.

Be ready to share and discuss your responses to each of the following questions with your group.

1. What has been your own experience with feeling an airline is reliable or unreliable?
2. What were some of the factors that *Forbes Magazine* considered about each airline as a part of their reliability rankings for the discount air carriers?
3. Are there other factors you might want to consider when judging an airline's reliability?
4. Would you consider all factors equally or would you rate some factors higher than others? Explain.

COURSE ACTIVITY: ARRIVAL DELAY TIMES

Share and discuss your responses to each of the following questions with your group.

20. What has been your own experience with feeling an airline is reliable or unreliable?

   Top priority would be for the airline to leave on time. Secondly would be proper handling of bagage.

21. What were some of the factors that *Forbes Magazine* considered about each airline as a part of their reliability rankings for the discount air carriers?

   1. Cancellations
   2. Delays
   3. Baggage Loss
   4. Complaints

22. Are there other factors you might want to consider when judging an airline's reliability?

   Possibly their ability to provide a quality in flight experience. Also the price of the tickets.

23. Would you consider all factors equally or would you rate some factors higher than others? Explain.

I would not weight in an attempt to balance out the subjectivity.

GROUP TASK

Chicago Magazine wants to write a story about the experiences of travelers who fly from Chicago to Minneapolis. They have heard complaints about arrival delays for each of the airlines that depart from Chicago and fly to Minneapolis. The magazine wants to decide whether the airlines can be considered to be equally unreliable or if any airline is doing a better job in getting their passengers to their destinations on time. The editor-in-chief of the magazine has contacted your group to help provide information for their article. She wants to focus on two regional airlines, Mesa and American Eagle, that fly out of Chicago to Minneapolis. The editor has obtained flight arrival delay time data, based on flights that departed from Chicago and flew to Minneapolis in 2008, for your group to analyze in order to address the following questions:

- Is there a difference in the reliability as measured by arrival time delays for these two regional airlines out of Chicago to Minneapolis? Or are both airlines pretty much the same in terms of their arrival time delays?
- Are any differences you find large enough to influence travelers so that they are advised to choose one airline over the other (all other factors, like cost, being equal)?

EXPLORE AND DESCRIBE

Examine the arrival time delays for the two airlines using the first set of data. The data can include negative numbers—which indicate the number of minutes that an airline arrived earlier than its scheduled arrival time.

Come up with *at least three numerical measures* that can be used to measure and compare the two airlines' reliability based on these data. Compute these measures for each airline and describe in words what each of your measures would indicate about an airline's reliability. These descriptions should be written so that a traveler can easily understand what they are measuring.

In the comparison of two seperate airlines, the more reliable airline is found to have a avg arrival time 25 min earlier than the less reliable, and have the majority of flights arrive in a time period c/o min smaller than the less reliable airline

DEVELOP RULES

For each one of your measures, compute the difference between the two airlines and decide how large the difference between the two airlines would need to be to say that one airline is truly more/less reliable than the other.

Use at least TWO of your measures to *develop and state a single rule* that another person can apply to data from two airlines to determine whether or not these airlines truly differ in their reliability.

However, an airline can be found to be reliable regardless, if its avg arrival time is earlier than 10 min late and the majority of flights arrive in a 30 minute window.

TEST RULES

Using four other sets of data collected from these same two airlines, apply your rule to each set of data. Adapt or modify your group's rule as you need to. This may include small changes or adding to or changing the measures that you used in your initial rule.

SUMMARIZE

Your group will now write a letter to the editor-in-chief of *Chicago Magazine* that includes the following:

- Your group's set rule, used to analyze the five sets of data. In your letter the rule needs to be clearly stated so that another person could apply them to compare data for two other airlines;
- A response to the editor's initial questions:
  - ❖ Is there a difference in overall reliability as measured by arrival time delays for these two regional airlines out of Chicago to Minneapolis? Or are both airlines pretty much the same in terms of their arrival time delays?
  - ❖ Are any differences you find large enough to influence travelers so that they are advised to choose one airline over the other (all other factors being equal)?

Type the letter in a word-processed document and email it to each of your group members and the instructor.

DISCUSSION

As a group, discuss your responses to each of the following questions.

24. Was it difficult to come up with a rule to determine whether one airline was really different in overall reliability than another airline? Why or why not?

25. How might your rules change if you were comparing more than two airlines for each city?

26. How might your rules change if you were comparing two airlines for three cities?

COURSE ACTIVITY: MEMORIZATION

Many times during the semester, you may feel like your brain just cannot hold all of the information you are learning in classes. Are there ways to improve our memories so that we can comprehend even more information? Research in cognitive psychology has suggested that the answer to that question is a resounding "yes". This literature has suggested several strategies to improve memory, enhance recall and increase retention of information.

One of the strategies identified by cognitive psychologists is that of chunking. Chunking refers to the process of taking individual units of information and grouping them into larger units (chunks). One common example of chunking occurs when we write and recall phone numbers. For example, a sequence of digits in a phone number, say 8-6-7-5-3-0-9, would be chunked into 867-5309.

In this activity, you will be exploring the following research question:

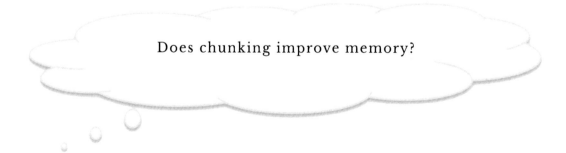

Does chunking improve memory?

To examine this research question, you will use the data collected from the memory experiment your class just partook in.

EXAMINING THE OBSERVED DATA}

The first part of any analysis is to examine the observed data. These are the data that are observed in the research study. Before you can examine the data, however, you will need to enter the data collected in the study into a TinkerPlot™ case table. See instructions below for how to set up the table.

> **Setting up the Case Table**
> - Drag a `Table` from the object toolbar into your document.
> - Create a new attribute called *Score* in the first column of the case table.
> - Create another new attribute called *Condition* in the second column of the case table.

Each row in the table will comprise a subject in the research study. Each column will comprise an attribute of the subject. For our purposes, you will need to enter data for two attributes. The first attribute will indicate the subject's score (i.e., the number of letters recalled) from the memory experiment. This is called the **response variable** since it contains data on the subjects' responses to the experiment. The second attribute will indicate the treatment condition that the subject was assigned to. This is called a **treatment variable**. In this research study the two treatment conditions are the **experimental condition** (chunking) and the **control condition** (no chunking).

- Enter the observed data from your class experiment into a TinkerPlots™ case table.
- Plot the observed data (see instructions on next page).

> **Plotting the Observed Data**
> - Drag a `Plot` from the object toolbar into your document.
> - Drag the response variable from the case table to the x-axis of the plot.
> - Drag a case icon to the right until the cases in the plot are fully separated (e.g., no vertical bin lines). You can also double-click on one of the endpoints and change `Bin width = 0`.
> - Drag the treatment variable from the case table to the y-axis of the plot.
> - Click the `Vertical Stack` button in the upper plot toolbar to organize the icons.

1. Sketch the plot below that you just created in TinkerPlots™.

SUMMARIZING THE DIFFERENCE BETWEEN THE TWO CONDITIONS

In order to answer the research question, you need to summarize the difference between the treatment and control conditions into a single number. You typically do this by finding the mean for each condition, and then computing the difference between the means. The difference in means satisfies the need for a single number summary. It also has another very nice quality, and that is the difference in means is interpretable. The difference in means indicates *how much better* the typical subject in the experimental condition does than a typical subject in the control condition.

2. Find the difference in means for the observed data by subtracting the mean score for the control condition from the mean score for the experimental condition.

3. Interpret this difference using the context of the memory study.

4. If there is *not* an effect of chunking on memory, what would you expect the difference in means to be? Explain.

5. Does the difference you found in the observed data *suggest there is an effect* of chunking on memory or not? Explain.

CONSIDERING CHANCE VARIATION AS AN EXPLANATION FOR THE DIFFERENCE IN MEANS

Before you conclude that chunking has an effect on memory, consider another alternative: *the difference in means you saw in the observed data is solely attributable to chance variation.* Think back to the activities and homework you have previously completed. Would you conclude that a coin is "unfair" because you flipped it ten times and got six heads? Probably not.

If you repeatedly flip a coin ten times, sometimes you get five heads. Other times you get four heads, or three heads, or seven heads, etc. This variation in the results is not because you flipped a different coin or used a different model, but rather because **the random process introduces variation into the results**.

Concluding that there is an effect of chunking on memory just because the difference in means from the observed data is not zero is akin to concluding that a coin is not fair just because you did not get exactly 50% heads!

How did you know whether a particular result—say 80 heads in 100 flips— indicated a coin was unfair? You modeled *flipping a fair coin* 100 times. By examining the variation in the results of repeatedly carrying out this experiment, you could investigate whether the observed result of 80 heads was likely or not under the assumption of flipping a fair coin.

You will use the same ideas to examine the result obtained from our observed memory data.

THE "JUST-BY-CHANCE" MODEL

To examine whether a result obtained in the observed data is solely due to chance (i.e., all the variation is due to the random assignment), one approach is to imagine the *scenario under which the chunking had no effect,* whatsoever. Under this assumption or scenario, evidence would be collected to determine if the difference in means that was observed in the data is too large to probabilistically believe that there is no effect of chunking. This statement or assumption of no effect of chunking is called the **null hypothesis** and is written as,

> H_0: There is no difference in the mean number of letters recalled between the control and experimental conditions.

If chunking is truly ineffective, then each subject's score on the memory test is only a function of that person and not a function of anything systematic, such as the chunking. The implication of this is that, had a subject been assigned to the other condition (through a different random assignment), her score on the memory test would have been identical since, in a sense, both conditions are doing nothing in terms of affecting the memory test scores.

RE-RANDOMIZATION: INSPECTING OTHER POSSIBLE RANDOM ASSIGNMENTS OF THE SUBJECTS

A researcher can take advantage of the idea that each subject's score on the memory test would be identical whether she was assigned to treatment or control and examine other possible random assignments of the subjects to conditions that could have occurred. To do this,

- In TinkerPlots™ drag a new `Case Table` into your document.
- Copy and paste the subjects' observed scores on the memory test (i.e., the response attribute) from the previous case table into this new case table. Call this attribute *Scores*.
- Create a new attribute in the case table called *RA1* (for Random Assignment 1).
- Add three more attributes to the case table, labeling them *RA2*, *RA3*, and *RA4*.

The new case table should now have an attribute of the original observed memory test scores and an empty attribute where you will input the "new" random assignment of conditions.

	Score	RA1	RA2	RA3	RA4
1	16				
2	10				
3	21				
4	8				
5	4				
6	3				
7	3				
8	9				

Screenshot of the newly created case table.

PHYSICAL SIMULATION OF THE RE-RANDOMIZATION

To aid you in creating these "new" random assignments of conditions, fill in the following:

> In the original experiment, _____ subjects were randomly assigned to the experimental (chunking) condition and _____ subjects were assigned to the control (no chunking) condition.

- You will be given several index cards with either an *E* (for experimental) or a *C* (for control). Each index card represents a single subject. Count the index cards to be sure that you have the same number of *E* cards as subjects originally assigned to the experimental condition and the same number of *C* cards as subjects originally assigned to the control condition.
- Shuffle the index cards together several times.
- Deal the shuffled index cards out one at a time. Record the condition on the first index card in the *RA1* attribute so it corresponds to the first subject. Continue recording the condition on each subsequent index card in turn for each subsequent subject.
- Plot the re-randomized data. (If you have forgotten how to plot the data, go back and re-read the directions from earlier in this activity.)

The data in the case table, and in the plot you just created, represent another way that the subjects could have been randomly assigned to the two conditions. This random assignment likely has different subjects in the control and experimental conditions than the observed data. Because of this, the mean memory score for the two conditions will also likely differ from the observed data. This, in turn, implies that the difference in means will also be different.

6. Compute and record the **difference in means** for the re-randomized (RA1) data. Be sure that the order you use when subtracting is consistent with the order you subtracted to obtain the original observed result. (Note: You may obtain a negative number here.)

- Repeat the re-randomization process three more times. Each time, record the data into TinkerPlots™, compute and record the difference between the means of the two groups. (Remember to subtract the mean score for the control condition from the mean score for the experimental condition.)
- Input each of the four differences you obtained from the four re-randomizations into your instructor's computer.

EXAMINING THE DISTRIBUTION OF THE DIFFERENCE IN MEANS

7. Sketch the plot of the difference in means (shown on your instructor's computer) below.

8. Does it look like it centers around zero? Explain why the distribution is centered at zero. (Hint: Think back to what the null hypothesis was.)

9. Quantify the level of support for the "just-by-chance" model based on the observed result (i.e., report the *p*-value)

10. What does this suggest about whether the difference in means that was observed is solely due to chance? Explain.

11. Use your answer from Question 10 to provide an answer the research question.

Overview of the Inferential Process for Comparing the Two Conditions in the Memorization Experiment

If there really were no effect of the grouping of letters, is it possible that random chance alone could have resulted in such an extreme observed difference between the two conditions? Once again, the answer is yes, this is indeed possible. Also once again, the key question is *how likely would it be for random chance alone to produce experimental data that favor the chunking condition by at least as much as the observational data do*. You will aim to answer that question using the following simulation analysis strategy:

- **Model:** Assume that there is no effect of the grouping of letters on the scores (the "just-by-chance" model).

- **Simulate:** Replicate the random assignment of these subjects and their memory scores between the two conditions. You will repeat this random assignment a large number of times. Each time you will calculate a measure of how different the conditions are, in order to get a sense for what is expected and what is surprising.

- **Evaluate:** Using the observed result, evaluate the level of support the data offer to the hypothesized "just-by-chance" model.

SLEEP DEPRIVATION READINESS

To introduce you to the context of the data you will be using in the next course activity, we would like you to read an article describing a study that was reported in *Nature*.

- Read the article, *Visual Discrimination Learning Requires Sleep after Training*. The article is available at http://www.nature.com/neuro/journal/v3/n12/pdf/nn1200_1237.pdf.

COURSE ACTIVITY: SLEEP DEPRIVATION

Sleep deprivation has been shown to have harmful effects such as fatigue, daytime sleepiness, clumsiness and weight loss or weight gain. Researchers have also established that sleep deprivation has a harmful effect on learning. But do these effects linger for several days, or can a person "make up" for sleep deprivation by getting a full night's sleep in subsequent nights?

Stickgold, James, and Hobson (2000), in a recent study, investigated this question by randomly assigning 21 subjects (volunteers between the ages of 18 and 25) to one of two groups: One group was deprived of sleep on the night following training and pre-testing with a visual discrimination task, and the other group was permitted unrestricted sleep on that first night. Both groups were then allowed as much sleep as they wanted on the following two nights. All subjects were then re-tested on the third day.

In this activity, you will be exploring the following research question:

> Does the effect of sleep deprivation last, or can a person "make up" for sleep deprivation by getting a full night's sleep in subsequent nights?

Subjects' performance on the test was recorded as the minimum time (in milliseconds) between stimuli appearing on a computer screen for which they could accurately report what they had seen on the screen. The sorted data and plots presented here are the improvements in those reporting times between the pre-test and post-test (a negative value indicates a decrease in performance):

Sleep Deprived ($n = 11$)	Unrestricted Sleep ($n = 10$)
−14.7	−7.0
−10.7	11.6
−10.7	12.1
2.2	12.6
2.4	14.5
4.5	18.6
7.2	25.2
9.6	30.5
10.0	34.5
21.3	45.6
21.8	

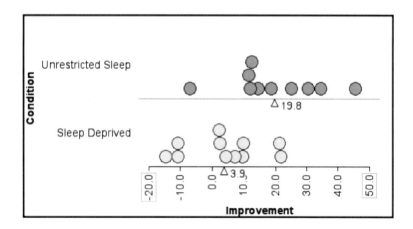

Observed data and plot of the observed data for the sleep deprivation study. The triangle under each plot indicates the mean improvement score for the respective group.

Discuss the following questions.

1. Does it appear that subjects who got unrestricted sleep on the first night tended to have higher improvement scores than subjects who were sleep deprived on the first night? Explain briefly.

2. Is the mean improvement higher for those who got unrestricted sleep? Calculate the difference in the means of the improvement scores. Does this appear to be a large difference?

3. Is it possible that there is really no harmful effect of sleep deprivation, and random chance alone produced the observed differences between these two groups?

In this study, the random chance is introduced not through the sampling process (like in Unit 1), but rather in the random assignment to groups. While it is possible that sleep deprivation has a harmful effect, it is also possible that it does not have a harmful effect and the researchers were just unlucky and happened to "assign" the subjects who were going to have more improvement in their test scores (regardless of which group they were going to be assigned to) into the unrestricted sleep group.

Similar to the simulation study in the *Memorization* course activity, one way to examine this is to consider what you would likely see if there really is no difference in test score improvement between the two conditions. (This is the null hypothesis!) If that is the case, it is reasonable to assume that these subjects would improve the same amount regardless of which group they had been assigned to because the effect on test scores would be identical for both groups.

Even if that is the case, however, since there are many possible ways to randomly assign 21 subjects into two groups, it is possible that the random assignment that came up was just unlucky and happened to "assign" the subjects who were going to have more improvement in their test scores into the unrestricted sleep group. If the random assignment had come up differently, it might have appeared as though there were no harmful effects, or even that the effects were beneficial—all because of random chance!

> **The key statistical question is:** If there really is *no difference between the conditions* in their effects on test improvement, how unlikely is it to see a result as extreme or more extreme than the one you observed in the data just because of the random assignment process alone?

MODELING THE DIFFERENCES IN IMPROVEMENT UNDER THE "JUST-BY-CHANCE" MODEL

You will conduct a randomization test using TinkerPlots™ to find out how likely it would be, *assuming there is no difference* between the conditions in their effects on test improvement, to see a result as extreme or more extreme than the mean difference you observed in the data (mean difference of 15.9) just because of the random assignment process alone.

The underlying idea of the randomization test is that the experimental condition labels are re-randomized—which produces a different random assignment of the subjects to the two conditions that could have occurred. The difference in the mean improvement scores is then computed under this re-randomization. This process of re-randomizing of the data and computing the difference in the mean improvement scores is repeated many times. The distribution of these differences displays the variation expected just because of the random assignment process alone and the observed result can be used to evaluate the "just-by-chance" model.

RANDOMIZATION TESTS IN TINKERPLOTS™

In order to carry out a randomization test using TinkerPlots™, you need to include multiple devices in the sampler. The first device will include the observed response data for all of the subjects. The second device will contain the experimental conditions.

Modeling a Set of Fixed Responses Under the "Just-by-Chance" Model

Under the null hypothesis of no difference between the two experimental conditions, the response values for the subjects are fixed—they will always be the same for the subjects, regardless of which experimental condition the subject is assigned. To produce simulated data that are fixed, you will use a sampling device called a `Counter`. Whereas the devices you have used thus far (spinners and mixers) select elements and values randomly, the counter selects elements in the same systematic order (the outcomes are perfectly predictable).

- Open the *Sleep-Deprivation.tp* data set.
- Set up a model that will produce the fixed responses for the subjects under the "just-by-chance" model (see instructions below).
- Run the model.

> **Setting Up the Model: Fixed Responses**
> - Drag a new `Sampler` from the object toolbar into your blank document.
> - The default device in the sampler is a `Mixer` with three elements. Remove all three of the elements by repeatedly clicking the `Remove Element` button (−) in the device toolbar at the bottom of the sampler.
> - Drag a `Counter` from the device toolbar at the bottom of the sampler onto the current device.
>
>
>
> - Right-click the attribute *Improvement* in the case table and select `Copy Attribute`.
> - Click the grey area in the sampler window (outside of where the numbers will go) and from the `Edit` menu select `Paste Cases`.
> - Change the name of the device from *Attr1* to *Improvement*.
> - Change the `Draw` value to *1*, and the `Repeat` value to *21*.

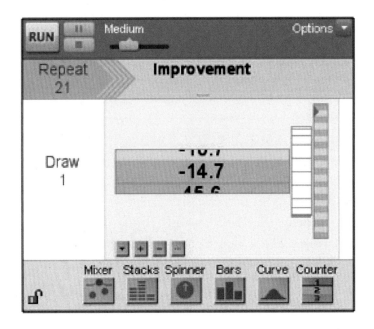

Screenshot showing the model to produce fixed responses from the counter.

4. Why did you change the Repeat value to *21*?

5. Run the model several times. Are you able to predict the outcomes generated in each trial?

Modeling the Random Assignment of the Treatment Condition Labels by Linking Multiple Devices

To model the random assignment of the treatment condition labels that might have occurred, you need to produce simulated data from another model that generates 11 labels of *Sleep Deprived* and 10 labels of *Unrestricted Sleep*. To do this you will use the `Stacks` sampling device.

You need to have the 21 fixed responses appear in the same case table as the 21 group labels. This allows us to easily attach a particular response to a label. To have the outcomes from multiple sampling devices appear in the same case table in TinkerPlots™, you *link* multiple sampling devices in the same sampler the same way you did in the *Pregnancy Test* course activity.

- Add a linked `Stacks` device to the sampler with the *Improvement* counter.
- Run the model.

> **Setting Up the Model: Randomly Assign Conditions**
> - Drag a `Stacks` sampling device from the device menu to the right-hand side of the existing *Improvement* counter. The sampler should contain two devices linked by a grey line.
> - Change the device name from *Attr2* to *Condition*.
> - Click the `Add Element` button (+) twice to add two elements to the stacks. These elements will indicate the treatment condition labels.
> - Change the label of the first bar from *a* to *Sleep Deprived*. Change the label of the second bar from *b* to *Unrestricted Sleep*.
> - Click on the `Device Options` button for the stacks device (upside-down triangle) and select `Show Count`.
> - Change the count value for the *Sleep Deprived* label to *11* and change the count value for the *Unrestricted Sleep* label to *10*.
> - Click on the `Device Options` button for the stacks device and from the `Replacement` menu select `Without Replacement`.

When you add linked devices, remember that the value for Draw changes automatically to the number of devices included in the sampler. A TinkerPlots™ sampler showing two linked devices modeling the random assignment of responses to conditions is shown below.

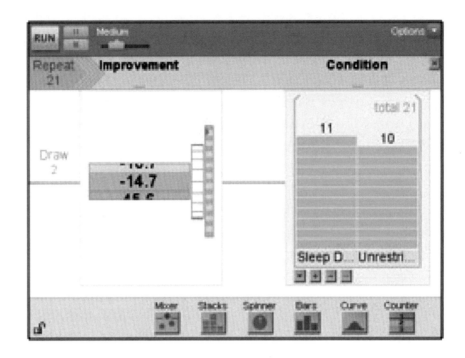

Screenshot showing the model to produce fixed responses from the counter and the random condition labels from the stacks.

The outcomes from both linked devices are recorded in the case table, each in their own attribute. In addition, an attribute called *Join* is also created that includes the outcomes of all the linked devices separated by a comma.

In this simulation, each trial represents *what might have occurred* under another random assignment of subjects to conditions.

- Plot the trial data. (Remember the response attribute from the trial's case table is dragged to the x-axis of the plot and the condition is dragged to the y-axis of the plot.)

6. Sketch the plot below.

As you have done in previous simulations, you will numerically summarize the trial results. From earlier in this activity, you summarized the observed data by computing the difference in the mean improvement scores between the two conditions. In fact, you computed,

$$\bar{X}_{\text{Unrestricted Sleep}} - \bar{X}_{\text{Sleep Deprived}}$$

You need to compute the summary of the trial data in the exact same manner.

- Use TinkerPlots™ to summarize the results of the trial by computing the mean improvement scores for each condition in the re-randomized data.

7. Compute the difference in means between the re-randomized groups in the trial.

COMPUTING THE DIFFERENCE IN MEANS

You need to have TinkerPlots™ compute the difference in means for the simulated data so that you can collect this result from many trials. The Ruler can be used to compute the difference between two values in a plot.

- Use the Ruler to compute the difference in means between the re-randomized groups in the trial (see instructions below).
- Check that the difference in means is the same as the difference you computed in the previous question. (If the difference calculated by TinkerPlots™ is correct, but has a reversed sign, you dragged the dotted lines to the wrong group.)

> **Computing the Difference in Means**
> - Click on the Ruler button in the upper plot toolbar.
>
>
>
> - Drag the right-most vertical dotted line on top of the mean-triangle for the *Unrestricted Sleep* condition. (Note that a purple circle will appear around the mean triangle.)
> - Drag the left-most vertical dotted line on top of the mean-triangle for the *Sleep Deprived* condition. (Note that a purple circle will appear around the mean triangle.)
> - The difference between the two means will be calculated in the lower left-hand corner of the plot window.

COLLECTING AND PLOTTING THE DIFFERENCE IN MEANS

- Collect the difference in means calculated by the ruler (which is reported in the bottom left corner of the plot).
- In the *History of Results* table, collect an additional 499 measures. (See instructions below for speeding up the simulation.)
- Plot the differences in means from the 500 simulated trials.

> **Speeding Up the Simulation**
> - Minimize all of the objects (sampler, results table, plot of the results) except for the collection window.
> - Select the collection window and from the Objects menu select Inspect Collection.
> - Uncheck the Animation On option.
> - Close the inspector window.
> - In the History of Results table, change the number of samples to collect to 499.
> - Click the Collect button.

8. Sketch the plot of the results (i.e., mean differences) from the 500 simulated trials below.

9. What are the cases in the plot? (Hint: Ask yourself what each individual dot represents.)

10. Where is the plot of the results centered (at which value)? Explain why this makes sense. (Hint: Think about what the hypothesis for the "just-by-chance" model is.)

11. Based on the plot, is the observed result found in the original data likely to have arisen solely from random assignment? Explain.

EVALUATING THE HYPOTHESIZED MODEL

12. Quantify the level of support for the "just-by-chance" model based on the observed result (i.e., report the p-value?)

13. In light of your answers to the previous two questions, would you say that the results that the researchers obtained provide strong evidence that the effect of sleep deprivation is harmful (i.e., that the "just-by-chance" model is not correct)? Or can a person "make up" for the lost sleep by getting a full night's rest on subsequent nights? Explain your reasoning based on your simulation results. Include a discussion of the purpose of the simulation process and what information it revealed to help you answer this research question.

EXAMPLE WRITE-UP FOR SLEEP DEPRIVATION STUDY

When reporting the results of a simulation study, pertinent information from the analysis that needs to be included is:

- **The type of test used in the analysis (including the number of trials):** Randomization test (500 trials)
- **The model assumed in the test:** The "just-by-chance" model which assumes any difference in the mean improvement scores between the two conditions is only due to random chance, not to effects of sleep.
- **The observed result based on the data:** Unrestricted sleep (M = 19.82 milliseconds), sleep deprived (M = 3.9 milliseconds), and the difference in means (15.92 milliseconds)
- **The p-value for the test:** p = 0.02 (one-tailed)
- **All appropriate inferences based on the p-value and study design:** Observed results do not support the "just-by-chance" model

As an example, we may write-up the analyses of the sleep deprivation study as follows:

> To study whether or not a person can "make up" for lost sleep by getting a full night of sleep on subsequent nights, 21 study participants were randomly assigned to one of two conditions. The subjects assigned to the sleep deprived condition ($n = 11$) were deprived of sleep on the night following training and pretesting with a visual discrimination task. The subjects assigned to the unrestricted sleep condition ($n = 10$) were permitted unrestricted sleep on that first night. Both groups were then allowed as much sleep as they wanted on the following two nights. All subjects were then retested on the third day. The data suggests that the group allowed unrestricted sleep ($M = 19.82$ ms) showed, on average, 15.92 ms more improvement than the sleep-deprived group ($M = 3.9$ ms).
>
> A randomization test was used to determine whether there was a statistically reliable difference in the improvement in reporting times between participants in these two groups. Assuming a "just-by-chance" model (no effect of sleep), a p-value of 0.02 (one-sided) was computed by re-randomizing the data 500 times. The observed difference of 15.92 ms provides strong evidence against the "just-by-chance" model. The results of this study may suggest that people cannot "make up" for sleep deprivation by getting a full night of sleep on subsequent nights.

COURSE ACTIVITY: LATINO ACHIEVEMENT

The *Center for Immigration Studies* at the *United States Census Bureau* has reported that despite shifts in the ethnic makeup of the immigrant population, Latin America—and Mexico specifically—remains this country's greatest source of immigrants. Although the average immigrant is approximately 40-years-old, large numbers are children who enroll in U.S. schools upon arrival. Their subsequent educational achievement affects not only their own economic prospects but also those of their families, communities and the nation as a whole.

Katherine Stamps and Stephanie Bohon studied the educational achievement of Latino immigrants by examining a random sample of the 2000 decennial Census data; a subset of which will be used in this activity[19]. One interesting research question that has emerged from their research is whether there is a link between where the immigrants originated and their subsequent educational achievement.

> Do immigrants from Mexico have lower average educational achievement scores than immigrants from other Latin American countries?

---

[19] Stamps, K., & Bohen, S. A. (2005). Educational attainment in new and established Latino metropolitan destinations. *Social Science Quarterly, 87*(5), 1225–1240.

The data in *Latino-Achievement.tp* contains a random sample of 150 Latino immigrants, of which 116 are from Mexico and 34 are from other Latin American countries. The response variable is an educational achievement score, ranging from 1 to 100, in which higher values indicate higher levels of educational achievement.

DISCUSS THE FOLLOWING QUESTIONS.

1. For this study, identify each of the levels/conditions for the treatment variable.

   Country of origin

2. For this study, identify the response variable and whether the response variable is categorical or quantitative in nature.

   Achievement score

   Quantitative

EXAMINE THE DATA

- Open the *Latino-Achievement.tp* data set.
- Plot the achievement scores conditioned on (grouped by) "treatment".

3. Sketch the plot.

4. Is the mean educational achievement level higher for immigrants from other Latin American countries? Calculate the difference in the means of the educational achievement levels for the observed data (i.e., report the observed result).

5. Immigration from Mexico is **not** *causal* to having a lower achievement score. What are other possible explanations that you can think of for the difference between the two means?

MODELING ACHIEVEMENT UNDER THE "JUST-BY-CHANCE" MODEL

You will conduct a randomization test using TinkerPlots™ to find out how likely it would be, assuming there is no difference between the two immigration groups in their average educational achievement level.

6. State the null hypothesis that defines the "just-by-chance" model to be used to simulate data in this investigation.

- Set up a model that will produce the **fixed responses** (achievement scores) for the subjects under the "just-by-chance" model.
- Add a linked `Stacks` device to re-randomize the **group labels**.
- Run the model.

PLOTTING AND COLLECTING THE RESULTS

- Use TinkerPlots™ to plot the results for the trial.
- Collect the difference in mean achievement level between the two groups using the `Ruler` tool.

7. What is the mean educational achievement level of immigrants from Mexico for this single simulated trial? What is the mean educational achievement level of immigrants from Other Latin American countries for this single simulated trial? What is the difference in means between these two groups?

SIMULATE RESULTS UNDER THE "JUST-BY-CHANCE" MODEL

- Carry out 500 randomized trials of the simulation in TinkerPlots™
- Collect and plot the differences in means for the 500 simulated trials.

8. Sketch the plot below.

9. What are the cases in the plot? (Hint: Ask yourself what each individual dot represents.)

10. Where is the plot of the results centered (at which value)? Explain, based on the null hypothesis, why this makes sense.

EVALUATING THE HYPOTHESIZED MODEL

11. Report the *p*-value (i.e., level of support for the hypothesized model) based on the observed result.

> In addition to quantifying the level of support for the hypothesized model, many researchers also provide a qualitative description of this evidence. While there are no hard-and-fast rules for gauging how strong the evidence is against the hypothesized model, the following guidelines can be used:
> - A *p*-value above 0.10 constitutes little to **no evidence** against the hypothesized model.
> - A *p*-value between 0.05 and 0.10 constitutes **borderline/weak evidence** against the hypothesized model.
> - A *p*-value between 0.025 and 0.05 constitutes **moderate evidence** against the hypothesized model.
> - A *p*-value between 0.001 and 0.025 constitutes **substantial/strong evidence** against the hypothesized model.
> - A *p*-value below 0.001 constitutes **overwhelming evidence** against the hypothesized model.

12. Based on the *p*-value, how strong would you consider the evidence against the hypothesized model?

13. Based on the *p*-value, provide an answer to the research question.

14. Write a brief memo in which you report the pertinent results from the analysis. (If you have forgotten which results you should report, see the *Sleep Deprivation* course activity.)

RANDOM ASSIGNMENT

Experimenters try to assign subjects to groups so that lurking and potentially confounding variables tend to balance out between the two groups. Random assignment is a method to create groups that are similar in all respects except for the treatment imposed. Because of this equivalence, if the response variable turns out to differ substantially between the groups, you can attribute that difference to the difference in treatment conditions. Because of this, using random assignment has the potential to allow researchers to establish a *cause-and-effect* relationship between the treatment and response variables.

To further help you understand how random assignment can be used to draw causal inferences, we would like you to read a short research report. We would also like you to read an excerpt from a research methods website that will further explain what it means for random assignment to create "identical" groups.

- Read the research report, *Random Assignment Evaluation Studies: A Guide for Out-of-School Time Program Practitioners*. The report is available at http://www.childtrends.org/wp-content/uploads/2008/01/Random-Assigment-Evaluations.pdf.

- Read the web excerpt, *Probabilistic Equivalence*. This excerpt is available at http://socialresearchmethods.net/kb/expequi.php.

COURSE ACTIVITY: STRENGTH SHOE®

The Strength Shoe® is a modified athletic shoe with a 4-cm platform attached to the front half of the sole. Its manufacturer claims that this shoe can increase a person's jumping ability. In this activity you will be examining the following question:

> How can you design a study to evaluate whether the manufacturer's claim about the Strength Shoe® is legitimate?

DISCUSS THE FOLLOWING QUESTIONS.

1. If your friend who wears strength shoes can jump much farther than another friend who wears ordinary shoes, would you consider that compelling evidence that strength shoes **really increase** jumping ability? Explain.

2. Now suppose that you take a random sample of individuals by randomly selecting them from the population. You observe who does and does not wear strength shoes, and then compare the two group's jumping ability. If, on average, the group who wears strength shoes can jump much farther than the group who wears ordinary shoes, would you consider that compelling evidence that strength shoes **really increase** jumping ability? Explain.

Both descriptions above lack compelling evidence to claim that people who wear the Strength Shoe® jump farther.

The evidence from the first situation is based on **anecdotal evidence**. Anecdotal evidence results from situations that come to mind easily and is of little value in scientific research. Much of the practice of statistics involves designing studies and collecting data so people do not have to rely on anecdotal evidence.

The problem with the evidence from the second situation is that you do not know whether or not the two groups might differ in more ways than simply one. For example, subjects who choose to wear the strength shoes could be more athletic to begin with than those who opt to wear the ordinary shoes. If one group is more athletic than the other, it could **confound** the results of the study.

When investigating whether or not one variable **causes an effect** on another, researchers seek to exert control by creating a comparison group and then assigning subjects to either the treatment group or the comparison group. An **experiment** is a study in which the experimenter actively imposes the treatment condition on the subjects. Ideally, the groups of subjects are identical in all respects other than the condition, so the researcher can then see the variable's direct effects on the response variable.

A 1993 study published in the *American Journal of Sports Medicine* investigated the Strength Shoe® claim using 12 intercollegiate track and field athletes as study participants[20]. Suppose you also want to investigate this claim, and you recruit 12 of your friends to serve as subjects. You plan to have six people wear a Strength Shoe® and the other six wear an ordinary shoe and then measure each group's jumping ability.

RANDOM ASSIGNMENT

Random assignment is the preferred method of assigning subjects to treatment conditions in an experiment. One characteristic of random assignment that makes it a good method of assigning subjects to conditions is that under random assignment, each subject has an equal chance (probability) of being assigned to any of the treatment conditions. In addition, there are several other benefits to using random assignment. You will explore the properties and benefits of random assignment in this activity.

> The word "randomization" is a synonym for random assignment.

---

[20] Cook, S. D., Schultz, G., Omey, M. L., Wolf, M. W., & Brunet, M. F. (1993). Development of lower leg strength and flexibility with the strength shoe. *American Journal of Sports Medicine, 21*, 445–448.

3. How might you use TinkerPlots™ to assign the following 12 subjects to two groups (ordinary shoe group and Strength Shoe® group)?

Jasmine	Mary	Antonio
Tong	John	Davieon
Andreas	Keyaina	Ringo
Ka Nong	Paul	George

CONFOUNDING VARIABLES

Two factors that might affect jumping distance are a person's sex and their height. In every study, there are potentially many factors (aside from the treatment) that may be related to the response variable and, in turn, affect the results of the study. Statisticians refer to these variables as **confounding variables**.

If we can consider what these factors might be before we collect our data, we can measure them. We might refer to these confounding variables as **observed confounding variables**.

While some confounding variables may be identified and controlled in a study, others may not be identified initially by the researcher. These unidentified variables (or **lurking variables**) may mislead researchers into thinking that a treatment is effective (or not effective), when in reality, all of the difference in the response variable is a function of differences in the confounding variable, not differences in the treatment.

One example of confounding is the study that found that people with larger ears tend to be better at math. Do larger ears have anything to do with math skills? Not really. The differences in math skills that was observed was related to the confounding variable of age. It turns out that in general, younger children—who also tend to have smaller ears—are not as good at math as older children and adults—who tend to have larger ears!

While this may seem like a silly example and the mistake easily avoidable; in practice erroneous results because of unobserved confounding variables are prevalent in every field. Even the smartest and experienced researchers will probably not identify all of the confounding factors related to differences in the response variable need to be controlled.

Luckily, it turns out that the key to controlling for *all* of these confounding variables (both observed and unobserved) is to use random assignment in forming experimental groups. For the remainder of this course activity, you will examine how random assignment "equalizes" not only the observed confounding variables (e.g., sex), but also unobserved confounding variables (factors we haven't yet thought of).

Next, we will see how random assignment helps us deal with observed confounding variables by using TinkerPlots™ to simulate randomly assigning each subject to an experimental condition many times to see what we can expect to happen in the long run. Note that we have not provided any data on how far participants will jump: we are trying to figure out: **what we can know about factors (aside from the treatment) that might affect participants' outcome/response**.

OBSERVED VARIABLE: SEX

We want to examine the proportion of females in each group to see if there are sex differences in the two groups that could be explaining the jumping differences we saw between the groups.

- Open the *Strength-Shoe.tp* TinkerPlots™ file.

Note that the model has already been set up for you; there is a `Counter` device with the study participants and a `Stacks` device that is randomly assigning the group that participant will be in.

- Press `Run` to record the results of a single random assignment.

4. Record the names assigned to each group in this table, along with their sex and heights in the table below.

Strength Shoe® Group			Ordinary Shoe Group		
Name	Sex	Height	Name	Sex	Height

- Plot the attributes Sex (y-axis) and Group (x-axis) in a single plot;
- Organize and separate the cases based on both attributes.
- Display the percentage for each group.

5. Calculate and report the **proportion of females** in each group. Also subtract these two proportions (taking the Strength Shoe® group's proportion minus the ordinary shoe group's proportion).

Proportion of females in Strength Shoe® Group:

Proportion of females in Ordinary Shoe Group:

Difference in proportions (Strength Shoe®– Ordinary Shoe):

This is just a single random assignment (trial). Is there a long-run pattern in sex differences? To observe any differences that might arise, we need to collect *two measures*: the percentage of females in the "Strength Shoe" group and the percentage of females in the "Normal Shoe" group.

- Right-click on the percentage female for the Strength Shoe® group and select `Collect Statistic`.
- Right-click on the percentage female for the Ordinary Shoe group and select `Collect Statistic`.
- Use the `Formula Editor` to compute the difference in the percentage of females between the two groups. (Note: Subtract the Ordinary Shoe group from the Strength Shoe® group.)
- Collect 499 more trials.
- Plot the 500 differences.
- Organize and fully separate the results (no bin lines) for the plot.
- Show the `Average` (and its numeric value) on both plots.

6. Sketch the plot below.

7. Where is this plot centered? What does this imply about the percentage of females in the two groups? Explain.

OBSERVED VARIABLE: HEIGHT

Now we want to examine the average height in each group to see if height differences in the two groups could be explaining the jumping differences we saw between the groups.

- Plot the attributes Height (y-axis) and Group (x-axis) in a single plot;
- Organize and separate the cases based on both attributes.
- Display the average for each group.

8. Calculate the average height for each group. Also find the difference in these two averages (taking the Strength Shoe® group's average minus the ordinary shoe group's average).

   Average height in Strength Shoe® Group:

   Average height in Ordinary Shoe Group:

   Difference in average height (Strength Shoe®− Ordinary Shoe):

Again, this is just a single random assignment and we want to get a sense of the difference in the average height across many random assignments.

- Right-click on the average height for the Strength Shoe® group and select `Collect Statistic`.
- Right-click on the average height for the Ordinary Shoe group and select `Collect Statistic`.
- Use the `Formula Editor` to compute the difference in the average heights between the two groups. (Note: Subtract the Ordinary Shoe group from the Strength Shoe® group.)
- Collect 499 more trials.
- Plot the 500 differences.
- Organize and fully separate the results (no bin lines) for the plot.
- Show the `Average` (and its numeric value) on both plots.

9. Sketch the plot below.

10. Where is this plot centered? What does this imply about the percentage of females in the two groups? Explain.

UNOBSERVED CONFOUNDING VARIABLES

Although it is great that random assignment will tend to "equalize" the observed variables (e.g., proportion of females or average heights) across groups, it would be simple enough to assign 1/2 of the females to the treatment, and 1/2 to the control group if we were concerned that the sex of a subject was related to the response variable.

But what about variables that might affect the response that we did not consider or even think about. Imagine that there is a genetic factor (which we did not measure before the study) that will strongly influence how much participants will improve from ANY training on their jumping ability (regardless of the shoe type). Let's call it the "*X-Factor*". This is a classic example of a unobserved confounding variable (or **lurking variable**): Since we do not know about it, we have no way to measure and control for it, but it will likely influence the results of our study. For example, what if more participants assigned to the Strength Shoe® group have this *X*-factor? Then the Strength Shoe® group would show increased jumping ability, even if training with a Strength Shoe is no better than training with an ordinary shoe.

- To explore this, we actually have an *X*-Factor already hidden in the TinkerPlots™ file! To show it, right click anywhere in the table of trial results and select `Show Hidden Attribute` (see screenshot).

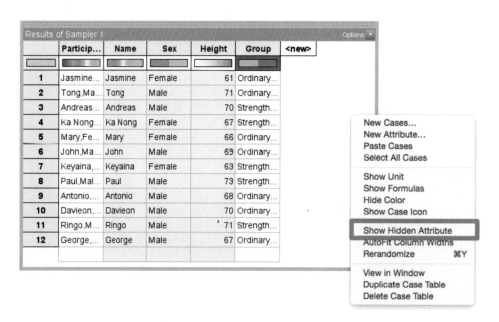

Screenshot of the `Show Hidden Attribute` *option after right-clicking on the trial results.*

You will now see that the *X*-Factor variable ("Yes" or "No", indicating whether the participant has that genetic factor or not) has appeared as another attribute in the trial results. It is important to remember that in real-life, we would not know about this confounding variable. But, here, we can examine how this confounding variable would be distributed between the Strength Shoe® and Ordinary Shoe groups when we use random assignment.

- Plot the attributes X-Factor and Group in a single plot.
- Separate the cases by the levels of the X-Factor.

11. What is the percentage of people with the *X*-Factor in the Strength Shoe® group? What is the percentage of people with the *X*-Factor in the Ordinary Shoe group?

Again, since this is only one trial, just like we did with males and females, we want to get a better sense of the differences in the *X*-Factor across many random assignments to groups. To observe any differences that might arise, we need to again collect *two measures*: the percentage of participants with the *X*-Factor in the Strength Shoe® group and the percentage of participants with the *X*-Factor in the Ordinary Shoe group.

- Right-click on the *X*-Factor percentage for the Strength Shoe® group and select Collect Statistic.
- Right-click on the *X*-Factor percentage for the Ordinary Shoe group and select Collect Statistic.
- Use the Formula Editor to compute the difference in the percentage of participants with the *X*-Factor between the two groups. (Note: Subtract the Ordinary Shoe group from the Strength Shoe® group.)
- Collect 499 more trials.
- Plot the 500 differences.
- Organize and fully separate the results (no bin lines) for the plot.
- Show the Average (and its numeric value) on both plots.

12. Sketch the plot below.

13. Where is this plot centered? What does this imply about the percentage of participants with the X-Factor in the two groups? Explain.

RANDOM SELECTION

During the *Latino Achievement* course activity you used a subset of Stamps and Bohen's data to examine the question of whether there was a difference in the educational achievement of immigrants from Mexico and that of immigrants from other Latin American countries.

HOW ARE STUDY PARTICIPANTS ASSIGNED TO GROUPS/CONDITIONS?

One major difference between the study described in the *Latino Achievement* course activity and the study described in the *Sleep Deprivation* course activity that you have learned about is how the study participants were assigned to groups/conditions. In the *Sleep Deprivation* course activity the study participants were randomly assigned to groups/conditions. In the *Latino Achievement* course activity the groups/conditions were not assigned at all—they were what are referred to as **intact groups**. Remember that how participants are assigned to groups/conditions directly impacts the inferences and conclusions that can be drawn. If the groups/conditions are randomly assigned, any effects we find can be causally attributed to the group differences.

HOW ARE STUDY PARTICIPANTS INITIALLY SELECTED TO BE A PART OF THE STUDY?

Another difference between theses to studies is how the study participants were initially selected to be a part of the study, regardless of which group they were in. The study participants used in the study described in the *Latino Achievement* course activity were **randomly sampled** from a larger population. In the *Sleep Deprivation* course activity the study participants were all volunteers, they were not randomly sampled from the larger population. **The use of random sampling also directly impacts the inferences and conclusions that can be drawn.**

STUDIES THAT USE RANDOM SAMPLING

The goal of studies that employ random sampling is very different than the goal of studies that employ random assignment. With studies that employ **random assignment**, the goal is to draw **cause-and-effect conclusions** about a particular treatment. We can draw cause-and-effect conclusions from studies that have employed random assignment, because the possibility of alternative explanations, other than the treatment, can be ruled out (recall the *Strength Shoe®* course activity).

In studies with **random sampling**, the goal is not to draw cause-and-effect conclusions about a treatment, but rather to **generalize a conclusion** that was found in the sample data to the broader population from which that sample was drawn.

Consider the study described in the *Sleep Deprivation* course activity. Here, the focus was on whether or not there was a difference in performance that could be directly attributed to deprivation in sleep (a causal inference). It does not, however, allow us to say who performance will be lower for. Is it all people in the United States? Only people between the ages of 18–25? People between the ages of 18–25 who live in Massachusetts?

In a study that employs random sampling, like the study described in the *Latino Achievement* course activity, any effect or differences that are statistically reliable can be generalized to the larger population that the study participants were sampled from. In the study described in the *Latino Achievement* course activity, this means the differences in achievement can be generalized to all Latino immigrants in the United States since the Census is a random sample (for all intents and purposes) of the population of the United States.

METHOD OF ANALYSIS

Regardless of whether a study employs random assignment or random sampling (or both), the method of analyzing the data is the same. We use the randomization test. The key is how we interpret any statistically reliable findings afterward. The inferences we make need to be based on the study design. Random assignment allows for cause-and-effect inferences, and random selection allows for generalization inferences.

COURSE ACTIVITY: SAMPLING

In statistics, estimation refers to the process by which one makes inferences about a population or model, based on information obtained from a sample. In practice, it is often impossible to examine every unit of the population, so data from a subset, or sample, of the population is examined instead. The sample data provides statisticians with the best estimate of the exact "truth" about the population. The "truth" one is searching for in the population is typically a summary measure such as the population mean or population percentage. Summary measures of a population are called **parameters**. The estimates of these values from sample data are referred to as **statistics**.

Consider taking a sample of ten students from this class for the purpose of estimating the average number of credits an EPSY 3264 student takes per term.

Share and discuss your responses to each of the following questions with your group.

1. Describe at least two different ways you can choose a sample of students.

2. Share your two methods of sampling from question 1 with another group. Compare and contrast the different methods. Are some better or worse than others? Why or why not?

3. Choose a sampling method that you think is good based on your answer to Question 2. Using this sampling method, would you expect two different samples of students to yield the same estimate for the average number of credits? Why or why not?

4. Using that same sampling method, would your estimate of the average number of credits be a good estimate of the true average number of credits taken by EPSY 3264 students in this class? Why or why not?

5. Using that same sampling method, would your estimate of the average number of credits be a good estimate of the true average number of credits taken by EPSY 3264 students in the last five years? Why or why not?

When estimating a parameter for an unknown model, there are several qualities that are ideal to have. Two of those qualities are **unbiasedness** and **precision**. Both of these qualities describe the estimation or sampling method used. You will examine unbiasedness in this activity and precision in upcoming activities.

UNBIASEDNESS

Unbiasedness is a quality that indicates that the estimation method used produces a distribution of the estimated parameter that is neither systematically too large nor too small. To illustrate this, consider the following two targets which show the locations of five darts thrown overhanded (target on the left) and underhanded (target on the right).

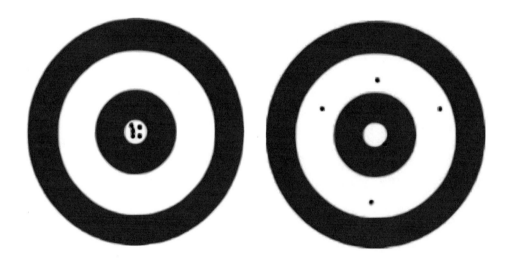

Both throwing methods, under- and overhanded, would be unbiased. If you examine the set of throws as a whole on the target on the left, they "average out" to be on center. Now examine the throws on the target on the right. Again, even though none of the darts thrown hit the center exactly, as a whole, the five darts "average out" to have "hit" the center. Now compare this with the targets below in which the darts were thrown under- and overhanded while the thrower had closed her eyes. (Not a good idea when throwing darts!)

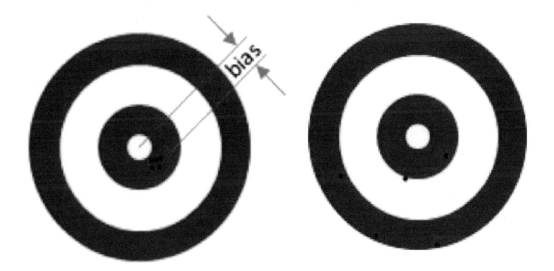

In both of these targets, the throwing method would be considered biased. On "average", the throws did not hit the center of the target. It is important to note that in examining the dart throwing methods, you used the distribution of throws to judge whether or not the throwing method was unbiased. Similarly, in judging whether an estimation or sampling method is unbiased, you will have to examine a distribution of the estimates produced using that method.

> Does the sampling method used impact whether the estimation is unbiased?

To help answer this research question, you are going to compare two different sampling methods using the population of 268 words in the passage on the following page. The passage is, of course, Lincoln's Gettysburg Address, given November 19, 1863 on the battlefield near Gettysburg, PA.

> Four score and seven years ago, our fathers brought forth upon this continent a new nation, conceived in liberty, and dedicated to the proposition that all men are created equal.
>
> Now we are engaged in a great civil war, testing whether that nation, or any nation so conceived and so dedicated, can long endure. We are met on a great battlefield of that war.
>
> We have come to dedicate a portion of that field as a final resting place for those who here gave their lives that that nation might live. It is altogether fitting and proper that we should do this.
>
> But, in a larger sense, we cannot dedicate, we cannot consecrate, we cannot hallow this ground. The brave men, living and dead, who struggled here have consecrated it, far above our poor power to add or detract. The world will little note, nor long remember, what we say here, but it can never forget what they did here.
>
> It is for us the living, rather, to be dedicated here to the unfinished work which they who fought here have thus far so nobly advanced. It is rather for us to be here dedicated to the great task remaining before us, that from these honored dead we take increased devotion to that cause for which they gave the last full measure of devotion, that we here highly resolve that these dead shall not have died in vain, that this nation, under God, shall have a new birth of freedom, and that government of the people, by the people, for the people, shall not perish from the earth.

The goal in many studies is to provide information about some characteristic of a population. For example, you may want to say something about the percentage of Americans who would support a particular piece of legislation. Or, you may want to provide information about the average amount of time University of Minnesota students take to graduate. One potential solution to obtain such information would be to collect the necessary data from every member of the target population.

In many studies, however, it may not be feasible given time and money constraints to collect data from each member of the population. In these cases it is only possible to consider data collected for a smaller subset, or **sample** from that population. In these cases, the characteristic of the population would be estimated from the sample data and inferences would be drawn about the population. The key is then to carefully select the sample so that the results estimated from the sample are representative of the characteristic in the larger population.

> The **population** is the entire collection of who or what (e.g., the observational units) that you would like to draw inferences about. A **sample** is a subset of observational units from the population.

Circle ten words in the text of the Gettysburg Address such that the ten words you select constitute a representative sample (i.e., have the same characteristics) of the entire passage.

6. Describe how the ten words in your sample are representative of the 268 words in the population.

   I think these ten words get at the general idea that Abe is trying to put across.

7. Record the length (number of letters not including punctuation) for each of the ten words in your sample

   3   3   8
   6   8   5
   7   3
   5   11

8. Determine the average (mean) word length for your ten words. This sample average is an estimate of the average word length in the population.

Add your sample estimate to the case table on the instructor's computer.

9. Sketch the plot of all of the sample estimates. Make sure to label the axis appropriately.

10. The actual population average word length based on all 268 words is 4.3 letters. Where does this value fall in the above plot? Were most of the sample estimates around the population mean? Explain.

11. For how many groups in your class did the sample estimate **exceed the population average**? What proportion of the class is this?

12. Based on your answer to question 13, is the sample estimate just as likely to be above the population average as it is to be below the population average?

When the sampling method produces characteristics of the sample that systematically differ from those characteristics of the population, you say that the **sampling method is biased**. To try to eliminate potential biases, it is better to take a random sample. This should create a representative sample, no matter what variable is focused on. Humans are not very good "random samplers", so it is important to use other techniques to do the sampling for us.

SIMPLE RANDOM SAMPLING

A **simple random sample** (SRS) is a specific type of random sample. It gives every observational unit in the population the same chance of being selected. In fact, it gives every sample of size n the same chance of being selected. In this example you want every possible subset of ten words that could be sampled to have the same probability of being selected.

The first step in drawing a simple random sample is to obtain a **sampling frame** or a list of each member of the population. Then, you can use software to randomly select a sample from the sampling frame. We have already prepared a sampling frame of the words in the Gettysburg Adress for you and saved it in a Tinkerplots™ file.

Use Tinkerplots™ to Draw a SRS

- Open the file *Gettysburg.tp*.
- Draw a simple random sample of ten words from the sampler.

13. Record the ten randomly sampled words and their lengths:

Word	Length

Use TinkerPlots™ to automatically compute the length of each word in your sample. To do this,
- Create a new attribute in the case table called *wordLength*.
- Right-click the attribute name *wordLength* and select the `Formula Editor`.
- Select `stringLength()` from the `Text` functions, and add the sampled words attribute between the parentheses.

14. Use TinkerPlots™ to plot and compute the mean word length for your ten randomly sampled words. Record the mean below.

15. Use `Collect` to carry out a simulation to randomly sample ten words, compute their mean length 200 times. Sketch the plot of these means. Make sure to label the axis appropriately.

16. If the sampling method is unbiased the estimates of the population average should be centered "around" the population average word length of 4.3. Does simple random sampling produce an unbiased estimate of the population average? Explain.

SAMPLE SIZE

Even when an unbiased sampling method, such as simple random sampling, is used to select a sample, you do not expect the estimate from each individual sample drawn to match the population average exactly. You should see, however, that the estimates are just as likely to over- or underestimate the population parameter. Because of this predictability to the variation in the possible sample estimates, inferences drawn about the population are said to be valid.

On the other hand, if the sampling method is biased, any inferences made about the population based on a sample estimate may not be valid. In such cases the estimate of the parameter is more likely to be too large or too small compared to the parameter. It is therefore very important to determine how a sample was selected before believing inferences drawn from sample results.

Does changing the sample size impact whether the sample estimate is unbiased?

- Change the sample size from 10 to 25.
- Use TinkerPlots™ to draw 200 random samples of 25 words, and collect the average word length for each sample.

17. Sketch the plot of the sample estimates based on the 500 samples drawn. Make sure to label the axis appropriately.

18. Record the average value for the estimate of the average word length.

19. Does the sampling method still appear to be unbiased? Explain.

20. Compare and contrast the distribution of sample estimates for $n = 10$ and the distribution of sample estimates for $n = 25$. How are they the same? How are they different?

21. Using the evidence from your simulations, answer the research question: Does changing the sample size impact whether the sample estimates are unbiased?

POPULATION SIZE

It is clear that changing the size of the sample does not affect whether or not an unbiased estimate is produced. Now we examine another question:

> Does changing the population size impact whether the sample estimate is unbiased?

To examine this we will now sample from a population that is quadruple the size of the original population (size = 1072) while keeping the population characteristics the same (e.g., mean word length is still 4.3 letters).

- Open the file *GettysburgLargerPopulation.tp*.
- Draw a simple random sample of ten words from the sampler.
- Compute the word length for each randomly sampled word.
- Plot and compute the mean word length for the ten randomly sampled words.
- Collect the mean word length for 200 random samples.

22. Sketch the plot of the sample estimates based on the 200 samples drawn. Make sure to label the axis appropriately.

23. Record the average value for the estimate of the average word length.

24. Does the sampling method still appear to be unbiased? Explain.

25. Compare and contrast the distribution of sample estimates for $n = 10$ now that you are sampling from a larger population to the distribution of sample estimates for $n = 10$ from before. How are they the same? How are they different?

26. Use the evidence collected from the simulation to answer the research question: Does changing the size of the population impact whether the sample estimates are unbiased?

A rather counterintuitive, but very crucial, fact is that when determining whether or not a sample estimate produced is unbiased **the size of the population does not matter!** Even more counterintuitive might be that the precision of the sample estimate is unaffected by the size of the population! (You will learn about the precision of a sample estimate in Unit 3.) This is why organizations like Gallup can state poll results about the entire country based on samples of just 1,000–2,000 respondents as long as those respondents are randomly selected.

In summary, it is important to note some caveats about random sampling:

- One still gets the occasional "unlucky" sample whose results are not close to the population even with large sample sizes.
- Second, the sample size means little if the sampling method is biased. As an example, in 1936 the *Literary Digest* magazine had a huge sample of 2.4 million people, yet their predictions for the Presidential election did not come close to the truth about the population[21].
- The size of the population does not affect the bias of the estimate, even if a small sample size is used.

---

[21] See http://en.wikipedia.org/wiki/The_Literary_Digest\#Presidential_poll for mc about the sampling and the 1936 election.

COURSE ACTIVITY: DOLPHIN THERAPY

Swimming with dolphins can certainly be fun, but is it also therapeutic for patients suffering from clinical depression? To investigate this possibility, researchers recruited 30 subjects aged 18–65 with a clinical diagnosis of mild to moderate depression. Subjects were required to discontinue use of any antidepressant drugs or psychotherapy four weeks prior to the experiment, and throughout the experiment. These 30 subjects went to an island off the coast of Honduras, where they were randomly assigned to one of two treatment groups[22].

Both groups engaged in the same amount of swimming and snorkeling each day, but one group did so in the presence of bottlenose dolphins and the other group did not. At the end of two weeks, each subjects' level of depression was evaluated, as it had been at the beginning of the study, and it was determined whether they showed substantial improvement (reducing their level of depression) by the end of the study.

Is swimming with dolphins therapeutic for patients suffering from clinical depression?

---

[22] Antonioli, C., & Reveley, M. A. (2005). Randomised controlled trial of animal facilitated therapy with dolphins in the treatment of depression. *British Medical Journal, 331*, 1–4.

DISCUSS THE FOLLOWING QUESTIONS.

1. For this study, identify each of the levels/conditions for the treatment variable.

2. For this study, identify the response variable and whether the response variable is categorical or quantitative in nature.

The researchers found that 10 of 15 subjects in the dolphin therapy group showed substantial improvement, compared to three of 15 subjects in the control group.

3. Organize these data/results (i.e., frequencies) into a 2x2 table. This table is sometimes referred to as a **contingency table**.

	No Improvement	Improvement	Total
Control Group			
Dolphin Therapy Group			
Total			

4. What percentage of all 30 study participants improved?

5. What percentage of the 15 subjects assigned to the dolphin therapy condition improved?

6. What percentage of the 15 subjects assigned to the control condition improved?

7. Find the difference between the percentage of subjects assigned to dolphin therapy condition that improved and the percentage of subjects assigned to the control condition that improved.

8. Write a few sentences summarizing the results in the sample. This summary should include a summary of what the data suggest about: (1) the *overall improvement* of these depression subjects; (2) the differences between the two treatment groups; and (3) whether or not the data appear to support the claim that dolphin therapy is effective.

The above descriptive analysis tells us what you have learned about the 30 subjects in the study. But can you make any inferences beyond what happened in this study? Is it possible that there is no difference between the two treatments and that the difference in the observed data could have arisen just from chance variation inherent in the random assignment?

While it is possible that the dolphin therapy is effective, it is also possible that dolphin therapy is not more effective and the researchers were unlucky and happened to "assign" more of the subjects who were going to improve into the dolphin therapy group than the control group.

One way to examine this is to consider what you would likely see if 13 of the 30 people were going to improve (the number of subjects who improved in our sample) regardless of whether they swam with dolphins or not. If that is the case, you would have expected, on average, about six or seven of those subjects to end up in each group ("just-by-chance" suggests this).

> **The key statistical question is:** If there really is no difference between the therapeutic and control conditions in their effects of improvement, how unlikely is it to see a result as extreme or more extreme than the one you observed in the data just because of the random assignment process alone?

MODELING THE IMPROVEMENT UNDER THE "JUST-BY-CHANCE" MODEL

You will answer this question by using TinkerPlots™ to conduct a **randomization test** in order to replicate the results you could have gotten just because of the random assignment, *under the assumption that dolphin therapy is not effective*—the control and dolphin therapy conditions are equally ineffective at improving depression.

- Open the *Dolphin-Therapy.tp* data set.
- Set up a model that will produce the **fixed responses** for the subjects under the "just-by-chance" model using a `Counter`. (If you have forgotten how to do this, refer back to the instructions in the *Sleep Deprivation* course activity.)

- Add a linked `Stacks` device to **re-randomize the group/condition labels**. (If you have forgotten how to do this, refer back to the instructions in the *Sleep Deprivation* course activity.)
- Run the model.

PLOTTING AND COLLECTING THE RESULTS

- Use TinkerPlots™ to plot the results for the trial.
- Collect the results from the trial (see instructions below).

> **Plotting and Collecting Results**
> - Drag a new plot into the workspace and place the *Condition* attribute on the y-axis and the *Improvement* attribute on the x-axis.
> - Display the percentages for each cell.
> - In the plot window, right-click on the percentage value for the improved patients in the *Dolphin Therapy* group and select `Collect Statistic`.
> - Repeat for the *Control* group, again right-clicking on the percentage value for the improved patients.
> - Create a third attribute in your `History of Results` table and name it *Difference*.
> - Right-click *Difference* and select `Edit Formula`. Set up a formula that computes the percent difference between the two conditions.

9. What percentage of subjects assigned to the dolphin therapy condition improved? What percent of subjects assigned to the control therapy condition improved? What is the difference in percents between these two groups?

10. Run another trial of the simulation. What percentage of subjects assigned to the dolphin therapy condition improved in this trial? What percent of subjects assigned to the control therapy condition improved? What is the difference in percents between these two groups?

SIMULATE RESULTS UNDER THE "JUST-BY-CHANCE" MODEL

- Carry out 500 randomized trials of the simulation in TinkerPlots™
- Collect and plot the differences in percentages for the 500 simulated trials.

11. Sketch the plot below.

12. What are the cases in the plot? (Hint: Ask yourself what each individual dot represents.)

13. Where is the plot of the results centered (at which value)? Explain why this makes sense based on the null hypothesis.

EVALUATING THE HYPOTHESIZED MODEL

14. Report the p-value (i.e., level of support for the hypothesized model) based on the observed result.

15. Based on the p-value, provide an answer to the research question.

16. Can the researchers make cause-and-effect inferences and attribute the differences in improvement to the dolphin therapy? Explain based on the study design.

17. Are the differences in improvement generalizable to all people suffering from depression? Explain based on the study design.

OBSERVATIONAL STUDIES

In some studies, researchers do not assign study participants to groups/conditions—either randomly or not. One example of this is the study described in the *Latino Achievement* course activity. The groups, Mexican and Non-Mexican, were already intact—researchers did not assign them to groups. When the study participants are not assigned to conditions by a researcher the study is referred to as an **observational study**.

Although it may not be valid, typically researchers analyze data from an observational study using the same methods as they use for data from a study in which the study participants were assigned to groups—using the randomization test.

Again, the key is in the inferences you can make from any statistically reliable results. You typically **cannot draw cause-and-effect conclusions** from observational studies, because **the possibility of alternative explanations always exists**. For example, since being from Mexico (or not) was not randomly assigned, there are many potential explanations for the differences we found in achievement, aside from country of origin.

Observational studies may or may not incorporate random sampling. For example, the study described in the *Latino Achievement* course activity was an observational study (the conditions of Mexico and Other were not assigned by the researchers) that incorporated random sampling. Because of the random sampling, the researchers could generalize the difference in educational achievement to the larger population of Mexican and non-Mexican Latin American immigrants.

However, because the conditions were not randomly assigned (in fact, they were not assigned at all) it is not appropriate to make cause-and-effect statements attributing that difference solely to the intrinsic characteristic of country of origin. It is much more likely that there is another reason that explains these differences (e.g., socioeconomic differences, prior education, etc.).

In 1988, results released to the public from the *National Household Survey on Drug Abuse* created the false perception that crack cocaine smoking was related to ethnicity[23]. The analysis, which was based on observational data (researchers cannot assign race) showed that the rates of crack use among blacks and Hispanics were twice as high as among whites. The data were re-analyzed in 1992 by researchers from Johns Hopkins University to take into account social factors such as where the users lived and how easily the drug could be obtained. They found that after adjusting for these factors, there were no differences among blacks, Hispanics and whites in the use of crack cocaine.

To further help you understand the limitations of observational data, we would like you to read an excerpt from a research methods website.

- Read the web excerpt, *Observational Study*. This excerpt is available at http://www.experiment-resources.com/observational-study.html.

---

[23] Lillie-Blanton, M., Anthony, J. C., & Schuster, C. R. (1993). Probing the meaning of racial/ethnic group comparisons in crack cocaine smoking. *Journal of the American Medical Association, 269*(8), 993–997.

COURSE ACTIVITY: MURDEROUS NURSE

For several years in the 1990s, Kristen Gilbert worked as a nurse in the intensive care unit (ICU) of the Veteran's Administration hospital in Northampton, Massachusetts. Over the course of her time there, other nurses came to suspect that she was killing patients by injecting them with the heart stimulant epinephrine.

Part of the evidence against Gilbert was a statistical analysis of more than one thousand 8-hour shifts during the time Gilbert worked in the ICU[24]. Here are the data presented during her trial:

	Gilbert working on shift	Gilbert not working on Shift	Total
Death occurred on Shift	40	34	74
No death occurred on shift	217	1350	1567
Total	257	1384	1641

---

[24] Cobb, G. W., & Gehlbach, S. (2006). Statistics in the courtroom: United States vs. Kristen Gilbert. In R. Peck, G. Casella, G. Cobb, R. Hoerl, D. Nolan, R. Starbuck and H. Stern (Eds.), *Statistics: A guide to the unknown* (4th Edition), pp. 3–18. Duxbury: Belmont, CA.

You will use these data to answer the following research question:

> Were deaths more likely to occur on shifts when Kristen Gilbert was working than on shifts when she was not?

DISCUSS THE FOLLOWING QUESTIONS

1. Among all 1,641 shifts, what percentage of shifts had a death occur?

2. Among the 257 shifts when Gilbert was working, what percentage of shifts had a death occur?

3. Among the 1,384 shifts when Gilbert was not working, what percentage of shifts had a death occur?

4. Compute the difference between the percentage of shifts in which a death occurred when Gilbert was working and the percentage of shifts in which a death occurred when Gilbert was not working.

5. For this study, specify the treatment variable and each of the possible treatment levels.

6. For this study, specify the response variable and each of the possible response categories.

7. Were shifts that Gilbert was working more likely to have a death occur than on shifts when she was not?

8. Does the difference in percentages convince you that Gilbert was giving lethal injections of epinephrine to patients? Why or why not?

9. What might other possible explanations be for the difference between the two percentages?

10. Compare and contrast the study design (i.e., sampling and assignment to conditions), in this study to the study design used in the *Sleep Deprivation, Dolphin Therapy, Latino Achievement* studies.

MODELING THE CHANCE VARIATION UNDER THE ASSUMPTION OF NO DIFFERENCE

You will conduct a randomization test using TinkerPlots™ to find out how likely it would be, assuming there is no difference between the percent of shifts in which a death occurred when Gilbert was working and those in which she was not working.

- Open the *Murderous-Nurse.tp* data set.
- Set up a model that will produce the fixed responses (*Death*) for the "subjects" under the "just-by-chance" model.
- Add a linked `Stacks` device to re-randomize the group labels.
- Run the model.

PLOTTING AND COLLECTING THE RESULTS

- Use TinkerPlots™ to plot the results for the trial.
  - Drag a new plot into the workspace and place the *Death* attribute on the x-axis and the *Shift* attribute on the y-axis.
  - Vertically stack the points.
- Collect two results: (1) the percentage of deaths when Gilbert worked on a shift and (2) the percentage of deaths when Gilbert was not working a shift.
- Calculate the difference between the two percentages in the `History of Results` table.

SIMULATE RESULTS UNDER THE "JUST-BY-CHANCE" MODEL

- Carry out 500 randomized trials of the simulation in TinkerPlots™
- Plot the differences in means for the 500 simulated trials.

11. Sketch the plot below.

12. What are the cases in the plot? (Hint: Ask yourself what each individual dot represents.)

13. Where is the plot of the results centered (at which value)? Explain why this makes sense.

EVALUATING THE HYPOTHESIZED MODEL

14. Report the p-value (i.e., level of support for the hypothesized model) based on the observed result.

15. Based on the p-value, provide an answer to the research question.

16. Can we make cause-and-effect inferences and attribute the differences in death rate to the fact that Kristen Gilbert worked the shift? Explain based on the study design. If not, provide an alternative explanation for the difference in percentages.

17. Are the differences in death rate generalizable to the population of all 8-hour shifts at the hospital? Explain based on the study design.

LEARNING GOALS: UNIT 2

The activities, homework, and reading that you have completed for the second part of this course have introduced you to several more fundamental ideas in the discipline of statistics. The ideas and concepts you were introduced to in this unit form the basis for statistical methods for testing hypotheses about group comparisons and drawing appropriate inferences. In addition, you have learned how to use TinkerPlots™ to carry out a randomization test.

Below are the key concepts and skills from Unit 2 that you should have learned.

LITERACY/UNDERSTANDING (TERMS AND CONCEPTS)

You should understand the:

- Basic terminology related to comparing groups (e.g., categorical data, quantitative data, factor, treatment variable, response variable, experiment, observational study, random assignment, random sampling, etc.)
- Scope of conclusions/inferences that can be reached are based on the design of a study
    - Random assignment in drawing cause-and-effect conclusions/inferences
    - Random sampling in drawing generalizable conclusions/inferences
- "Just-by-chance" model is used to model the hypothesis of no effect of treatment/no group differences
- *P*-value is a quantification of the level of support for the "just-by-chance" model

- *P*-value is the proportion of results that are as extreme or more extreme than the observed result
- Desirable properties of estimators such as unbiasedness and precision
- Impact of sampling method on the bias and precision of estimates
- Size of the sample impacts the precision of an estimator, but not the bias
- Size of the population does not impact the precision of an estimator, nor the bias

MODEL-SIMULATE-EVALUATE FRAMEWORK

You should be able to:

- Use TinkerPlots™ to generate simulated data from the "just-by-chance" model that includes two sampling devices (one fixed, and one random) to produce the distribution of a numerical summary under the assumption of no effect of treatment or no group differences for randomly assigned or randomly sampled groups (distributions are centered at 0, no difference)

TINKERPLOTS™ SKILLS

You should also be able to do all of the following using TinkerPlots™:

- Create a sampler to model the random variation inherent in the assignment or sampling of two groups by
    - Using a `Counter` to systematically reproduce the fixed outcome or responses
    - Using a random device (e.g., `Stacks`) to randomly produce the two groups without replacement
- Create a plot that separates the trial outcomes for the two groups
- Use a formula to create a new attribute in a case table.
- Compute a numerical measure to summarize the difference between the two groups (e.g., difference in means)

In the next course activity, *Unit 2 Wrap-Up & Review*, you will have a chance to assess yourself on whether or not you have mastered these ideas through a variety of practice and extension problems. As a pre-cursor to this activity, you may want to review the readings and activities in Unit 2.

COURSE ACTIVITY: UNIT 2 WRAP-UP & REVIEW

TERMINOLOGY FOR UNIT 2

At this point, you should be familiar with the following terms. Write down what each term represents as well as any notes that may help you remember.

1. Experiment

2. Observational Study

3. Factor/Treatment Variable

4. Response Variable

5. "Just-by-Chance" Model to Compare Groups

6. Random Assignment

7. Random Sampling

8. Confounding Variables

9. Randomization Test

10. Bias

11. Population

12. Sample

13. For all studies described in Unit 2 (1) consider whether the study was an experiment or observational study; (2) identify the levels of the treatment/grouping variable and (3) identify whether the response variable was categorical or quantitative in nature.

Study/Activity	Type of study	Treatment/grouping variable	Response variable
Dolphin Therapy			
Latino Achievement			
Memorization			
Murderous Nurse			
Sleep Deprivation			
Strength Shoe®			

RATING CHAIN RESTAURANTS

The August 2012 issue of *Consumer Reports* (http://www.consumerreports.org/cro/magazine/2012/08/america-s-best-restaurant-chains/index.htm) included ratings of 102 chain restaurants. The ratings were based on surveys that readers sent in after eating at one of the restaurants. The article reported that the survey results were based on 110,517 visits to full-service restaurant chains between April 2011 and April 2012, and reflected the experiences of their readers, not necessarily those of the general population.

14. Do you think that the sample here was chosen randomly from the population of *Consumer Report* readers? Explain.

15. Why do the authors of the article make this disclaimer about not necessarily representing the general population?

16. To what population would you feel comfortable generalizing the results of this study? Explain.

NATIVE CALIFORNIANS

Out of people living in California, is the percentage born in California different in the years 1950 and 2000? To investigate this question, a random sample of 500 California residents was drawn using data from the 1950 Census and another random sample, independent from the 1950 random sample, of 500 California residents was drawn using data from the 2000 Census. The results are shown in the table below.

	1950	2000	Total
Born in California	219	258	477
Not born in California	281	242	523
Total	500	500	1,000

17. Compute the difference between the percentage of native California residents in the years 1950 and 2000.

18. Describe how to carry out a simulation to investigate the following research question: Has the percentage of native born Californians changed from 1950 to 2000?

19. Carry out the simulation using TinkerPlots™. Report all pertinent results of the simulation. Also report any appropriate inferences based on the study design.

BLOOD PRESSURE

In a 2001 study, volunteers with high blood pressure were randomly assigned to one of two groups. In the first group—the talking group—subjects were asked questions about their medical history in the minutes before their blood pressure was measured. In the second group—the counting group—subjects were asked to count aloud from one to 100 four times before their blood pressure was measured. The data presented here are the diastolic blood pressure (in mm Hg) for the two groups. The sample average diastolic blood pressure for the talking group was 107.25 mm Hg and for the counting group was 104.625 mm Hg.

Talking Group ($n = 8$)	Counting Group ($n = 8$)
103	98
109	108
107	108
110	101
111	109
106	106
112	102
100	105

20. Do the data in this study come from a randomized experiment or an observational study? Explain.

21. Calculate the mean difference between the two groups.

22. Write out the null hypothesis for this study.

23. Use TinkerPlots™ to carry out the appropriate analysis to determine if a difference this large could reasonably occur just by chance. Report all pertinent results of the simulation. Also report any appropriate inferences based on the study design.

SOCIAL FIBBING

A student investigated "social fibbing" (the tendency of subjects to give responses that they think the interviewer wants to hear) by asking students "Would you favor a policy to eliminate smoking from all buildings on campus?" She randomly assigned half the subjects to be questioned by an interviewer smoking a cigarette and the other half were interviewed by the same student but not while she was smoking. The results are displayed in the following table.

	Favor Ban	Do not Favor Ban	Total
Interviewer is Smoking	43	57	100
Interviewer is not Smoking	79	21	100
Total	122	78	200

Does the behavior of an interviewer affect the percentage of people who favor a smoking ban?

24. Calculate the observed result (mean difference) based on the research question.

25. Use TinkerPlots™ to carry out the appropriate analysis to determine if a difference this large could reasonably occur just by chance. Report all pertinent results of the simulation. Also report any appropriate inferences based on the study design.

UNIT III: STATISTICAL ESTIMATION

Far better an approximate answer to the right question...than an exact answer to the wrong question.

—John Tukey (1962)

One of the largest uses of statistical inference is the estimation of unknown parameters using sample data. Polling companies such as Gallup and Harris have made billions of dollars by using statistical estimation to carry out public opinion research. These companies are hired to provide snapshots of public attitudes and opinions on varied topics from politics and the economy, to social awareness and health and well-being. The results of their polls are seen on a daily basis in almost every newspaper, news blog and website the world over.

Statistical estimation is used by more than pollsters. Biologists, social scientists, and medical researchers use statistical sampling and estimation to quantify populations. For example, the Department of Natural Resources estimates the size of animal, bird, and fish populations yearly. These estimates are used to help set hunting and fishing regulations, as well as to allocate resources.

OUTLINE OF THE UNIT

In this unit, you will begin exploring ideas related to statistical estimation.

In the first activity, you will be introduced to three features of distributions that statisticians typically describe in an analysis of data. In the second activity, you will learn how to quantify variation in a distribution by computing the standard deviation. The standard deviation also plays a very important role in the third activity, in which you will quantify the precision of an estimator. This is important because it provides a numerical summary of the uncertainty in an estimate that is due to sampling, called sampling error.

After learning how to quantify the precision in an estimate, you will learn how statisticians use this measurement to provide a margin of error and interval estimates. You will also learn a method called the bootstrap, that will allow you to obtain estimates of the sampling error in order to estimate an unknown population parameter from a single observed sample of data.

The idea of the bootstrap in estimating the amount of sampling uncertainty will be developed further in an activity where you will also learn how to quantify the size of an effect, or effect size, between two groups. This is a natural extension of the ideas and group comparisons you experienced in the second unit. Here, you will learn how to answer the follow-up question to "are the groups different?", which is, "**how different** are the groups?"

Lastly, you will experience an activity in which you will gain a deeper understanding of how and why statisticians and researchers use the interval estimate produced by plus/minus (±) two standard errors. You will also learn about the connections between theoretical sampling from a population and empirical bootstrap sampling. Throughout the unit, you will continue to add to the knowledge you accumulated from Unit 1 and Unit 2, including both content knowledge and TinkerPlots™ knowledge.

REFERENCES

Tukey, J. (1962). The future of data analysis. *Annals of Mathematical Statistics* *33*(1), 1–67.

DESCRIBING DISTRIBUTIONS

One of the important steps in any statistical analysis is that of summarizing data. It is good practice to examine both a graphical and a numerical summarization of your data. These summarizations are often part of the evidence that researchers use to support any conclusions drawn from the data. They also allow researchers to discover structure that might have otherwise been overlooked in the raw data that was actually collected. Lastly, both graphical and numerical summaries of the data often point to other analyses that may be undertaken with the data.

Once raw data has been collected in a study, it can be overwhelming to pull any kind of meaning out of it. For example, it is not uncommon for Google to be dealing with millions of cases. How can Google—or any researcher for that matter—go from all of that raw data to something that can help them answer their research questions?

Rather than examining all of those cases individually, researchers examine the data collectively, often by plotting it. This is what is meant by a graphical summary of the data; it is quite literally, a picture of the distribution. For example, the plots of the exam scores you saw in the *Features of Distributions* course activity provided graphical summaries of the raw exam scores for each of nine classes.

There are many, many different types of plots that have been created to graphically summarize data. Each can provide a slightly different representation of the data. Metaphorically, you can imagine each of these different plot types as a different photo taken of the exact same person. Some may be color, others black and white. Some may be taken from different perspectives, angles or distances.

While all photographs "summarize" the same person, you may notice characteristics of that person in some photos that are not evident in others. Many of the photos, however, will show the same thing.

SHAPE

The dotplot[25] that TinkerPlots™ provides is a very useful plot. It allows us to summarize the **shape** of the distribution very easily. Shape is used to describe a distribution's symmetry. As you might expect, **symmetric distributions** are shaped the same on either side of the center. (Another way of thinking about this is that if you folded the distribution at the center, the folded half of the distribution would align pretty well on top of the other half.) For example, "bell-shaped" or "normal" distributions are symmetric.

When a distribution is asymmetric, it is referred to as a **skewed distribution**. The distribution shown in Figure 1 is a skewed distribution. In this distribution, there appears to be a longer tail on the right side of a distribution. Because the tail is on the right side of the distribution, statisticians would say it is "skewed to the right" or "positively skewed". In a similar way, a distribution that tails to the left is "skewed to the left" or "negatively skewed".

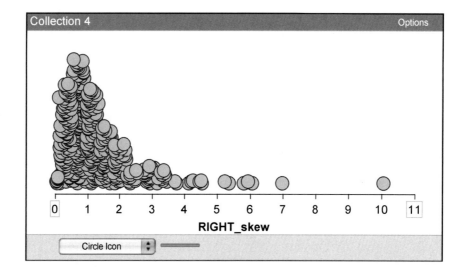

This distribution is skewed to the right, or positively skewed.

---

[25] TinkerPlots™ also provides other types of plots than the dotplot, including the box plot (sometimes called the box-and-whiskers plot) and the hat plot (a variation of the box plot).

LOCATION

Aside from the overall shape of the distribution, it is also useful to summarize the **location** of the distribution. The location of the distribution provides a summarization of a so-called "typical" value for the data. A "typical" value can be estimated from the plot of the distribution. You can also use more formally calculated summaries of the location such as the mean, median, or mode. These values are easily calculated using TinkerPlots™.

When looking at a plot of a distribution, data analysts often consider the number of modes or "humps" that are seen in a plot of the distribution. Here, the concept of mode is slightly different—although related—to the concept of mode that you may have learned in previous mathematics or statistics courses. The mode of a distribution gives a general sense of the values or measurements that occur frequently. This may be a single number, but many times is not. For example, the first hump of the distribution shown in the figure below suggests that values around nine are very common. The actual value of nine, however, may only show up once or twice in the data.

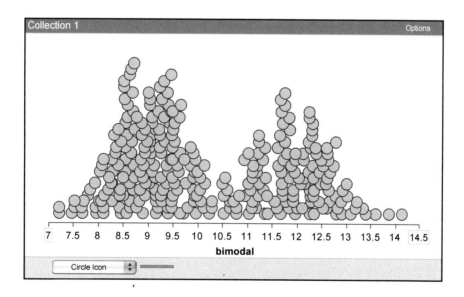

A bimodal distribution showing two modes. One mode is around nine, and the other is near 12.

A distribution can be **unimodal** (one mode), **bimodal** (two modes), **multimodal** (many modes), or **uniform** (no modes). The distribution shown above is bimodal—notice there are two humps. Uniform distributions have roughly the same frequency for all possible values (they look essentially flat) and thus have no modes.

VARIATION

The other characteristic of a distribution that should be summarized is the **variation**. Summarizing the variation gives an indication of how variable the data are. One method of numerically summarizing the variability in the data is to quantify how close the observations are relative to the "typical" value *on average*. Are the observations for the most part close to the "typical" value? Far away from the "typical" value? How close?

It turns out, that the shape of the distribution also helps describe the variation in the data. For example, "bell-shaped" distributions have most observations close to the typical value, and more extreme observations show up both below and above the typical value (the variation is the same on both sides of the "typical" value). Whereas skewed distributions have many observations near the typical value, but extreme values only deviate from this value in one direction (there is more variation in the data on one side of the "typical" value than the other).

One thing that affects the variation, and should be described is whether there are observations that stand out from the other observations. Often these observations have extremely large or small values relative to the other observations. These observations are referred to as **outliers**, or extreme cases. For example, in the positively skewed distribution shown previously, the observation that has a value near 10 would likely be considered an outlier.

PUTTING IT ALL TOGETHER

Rotten Tomatoes is a website which aggregates movie critics' reviews of films. The website marks each review as either positive or negative and then gives a score based on the percentage of positive reviews. In addition, the general public can also give a positive or negative review to a film. These reviews are tabulated and the score given to each film represents the percentage of positive reviews.

The following is a graphical summary of the scores for 134 movies released in 2009.

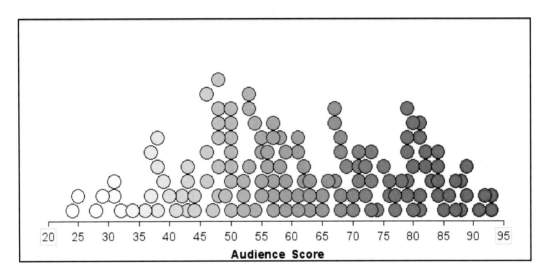

The scores for a sample of movies released in 2009 based on the general public's reviews. The scores represent the percentage of positive reviews for each movie.

A written description of the distribution might read as follows:

> The distribution of scores for this sample of movies is fairly symmetric. The median score for these movies is near 60, indicating that a typical movie released in 2009 is positively reviewed by about 60% of the public. The distribution also indicates that there is a lot of variation in the movies' scores. Most of the movies in the sample have a score between 35 and 85, suggesting large differences in the public's opinion of the quality of movies.

COURSE ACTIVITY: FEATURES OF DISTRIBUTIONS

Imagine multiple sections of the same college course, taught by different instructors. Below are a series of plots that depict the distributions of hypothetical exam scores in various sections.

1. Examine the three distributions of exam scores for classes *A*, *B*, and *C*. What are the primary differences between these three distributions? What are potential factors that might explain the differences?

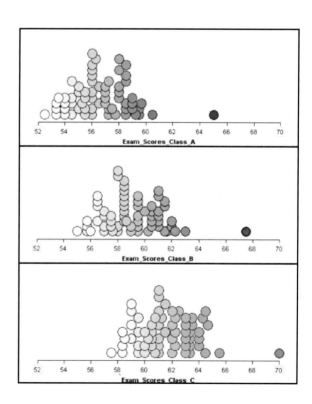

A has the overall worst scores and C has the overall best scores. This could result from the method of teaching in each class.

2. Examine the three distributions of exam scores for classes D, E, and F. What are the primary differences between these three distributions? What are potential factors that might explain the differences?

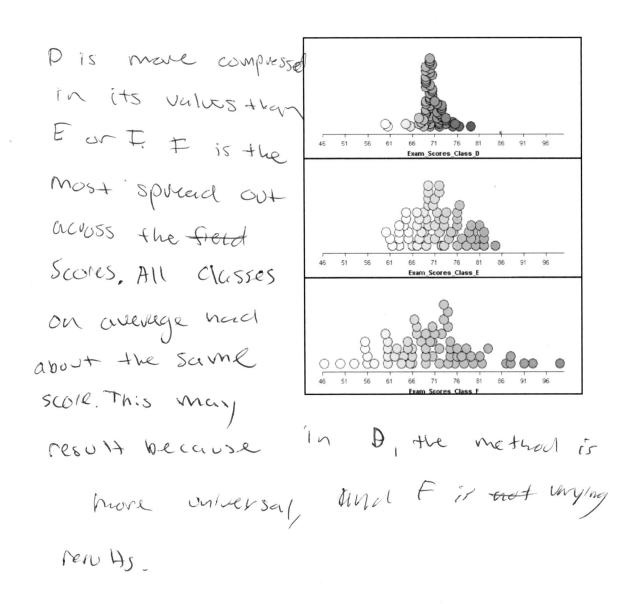

D is more compressed in its values than E or F. F is the most spread out across the field scores. All classes on average had about the same score. This may result because in D, the method is more universal, and F is ~~not~~ varying results.

3. Examine the three distributions of exam scores for classes G, H, and I. What are the primary differences between these three distributions? What are potential factors that might explain the differences?

They all inhabit the same area, but G's top score is 80 and trending down, while I's low score is 60, which then trends upward.

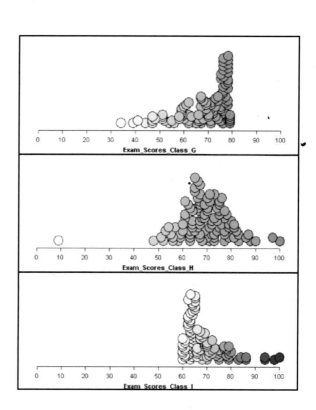

typical

CELL PHONE BILLS

Consider a survey study conducted on a random sample of 25 University of Minnesota students. One survey item asked students to self-report the amount of his or her last cell phone bill (in dollars). The plot of the bill amounts is shown below.

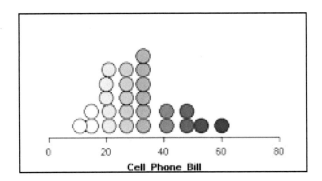

4. If you wanted to tell someone the amount of a "typical" cell phone bill for these students, what would you say?

   Somewhere around 30 dollars

5. How would you describe (quantify) the *overall* amount of variation in the distribution (i.e., for all 25 cell phone bills)?

   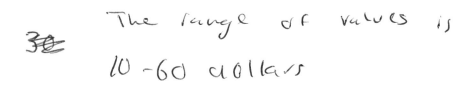

   The range of values is 10–60 dollars

6. How far from the typical amount that you identified in Question 4 are *most* of the students bills?

   30 dollars either way

7. What is a potential factor(s) that might explain the variation in these bills?

8. Using the typical cell phone bill you identified previously as a reference point, consider the amount of variation in the distribution on both sides of this point. Is the variation roughly the same on the left- and right-hand side of this point? Is there more or less variation on either side of this value?

NUMBER OF HOURS STUDIED

The plot below contains responses from 100 EPsy 3264 students who responded to the survey question: "How many hours per week do you typically study?" These students' responses are a random sample from all responses obtained from all classroom sections of EPsy 3264 taught from 2004–2010. Examine the plot of these data.

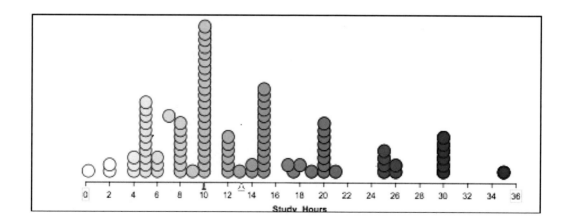

9. What does each dot (i.e., case) in the distribution represent?

10. Summarize the features of the distribution. Be sure to identify the **"typical"** amount of time spent studying and the **variation** in the amount of studying. (When describing the variation, you should quantify the "average" amount of deviation from the typical value.) You should also indicate the **shape** of the distribution.

11. What is a potential factor(s) that might explain the variation in these data?

UNDERSTANDING THE STANDARD DEVIATION

To help provide you with a deeper understanding of ideas related to the standard deviation, we would like you to complete a short online tutorial from *Usable Stats*.

- Complete the online, *Standard Deviation Tutorial*. The tutorial is available at http://www.usablestats.com/tutorials/StandardDeviation.

COURSE ACTIVITY: COMPARING HAND SPANS

In this activity, you will learn about the standard deviation, a common measure of variability.

> How can you quantify variability and summarize it into a single measure?

1. Measure and record the hand span for each person in your group.

2. Enter the data into a TinkerPlots™ case table. Create a plot of the hand spans for your group. Sketch the plot below. Be sure to appropriately label the x-axis.

3. Compute the mean hand span for your group using TinkerPlots™. Record the mean.

THE STANDARD DEVIATION

Recall that the mean is a single number that can be used to summarize the data. In this context, it is a description of the typical hand span measurement for your group. Of course, not every student in the sample is at the typical value (in fact all of them might be different from the typical value). Thus, it is also useful to have a single number description of how different the data tends to be from this typical value.

One single number description of the variability in a sample of data is called the **standard deviation** or *SD*. If the word "typical" is substituted for the word "standard" in its name, the name standard deviation (typical deviation) makes more sense. This measure quantifies variability by determining how far data cases typically deviate from the mean value.

- Use TinkerPlots™ to create a new attribute in the table, called *Deviations*, that contains the difference between the observed data (hand spans) and the mean of your group members' hand spans. Use a formula to compute this difference (you can compute these by subtracting the mean from each observation).
- Create a plot of the *Deviations* attribute.

4. How would you interpret the values of the *Deviations* attribute?

5. Sketch the plot below. Make sure to label the x-axis. Circle the mean and record its value.

6. How does the distribution of deviations compare with the distribution of hand span lengths you created in Question 2?

One thing you may not have known about the mean is that it is the value that "balances" the data. In other words, the mean is the value that gets the deviations to sum to zero. This is useful when describing a typical value of the data (it is the "closest" point to all of the cases, on average). If you try to average these deviations, however, you will always get zero. This is not very useful in summarizing variation in a data set, nor in comparing the variation between two data sets. One way to alleviate this problem is to square each of the deviations before you add them together.

- Create another attribute, *SquaredDeviations*, which contains the squared values of the deviations. (Again, use the `Formula Editor` to create this attribute.)
- Create a plot of the *SquaredDeviations* attribute.

7. Sketch the plot below. Make sure to label the x-axis.

8. How does the distribution of the squared deviations compare to the distributions of the original hand spans? How does it compare to the distribution of *deviations*?

9. Compute the mean of the squared distributions and record its value below. Because the deviations have been squared, this value represents the **typical squared deviation**.

10. Find the square root of the mean value and record its value below.

11. What does this new value represent (i.e., interpret its value)?

Computing the Standard Deviation using TinkerPlots™

Now use TinkerPlots™ to find the standard deviation of the original data directly.
- In the case table where you entered the original hand spans, create a new attribute called standardDeviation.
- Use the `Formula Editor` to compute the standard deviation of the hand spans by using the `stdDev()` function.

The value computed using TinkerPlots™ will be similar, albeit higher, than the value you obtained in Question 10. This is because there is a slight adjustment made to the denominator when a standard deviation is computed from a sample of data. From this point forward, you should always use the `stdDev()` function to compute the standard deviation.

USING BOTH THE MEAN AND STANDARD DEVIATION: A COMPLETE SUMMARY

The file *Study-Hours.tp* contains responses from 100 EPSY 3264 students who responded to the survey question: How many hours per week do you typically study? These students' responses are a random sample from all responses obtained from all in-class sections taught from 2004–2010.

12. Find and record the mean of these data.

13. Use the `stdDev()` function to find the standard deviation of these data. Record this value.

14. Use the mean and standard deviation to provide an interval estimate to answer the question: How many hours a week do students study?

15. Describe the **population of students** to which these sample estimates (mean and standard deviation) apply.

COURSE ACTIVITY: KISSING THE 'RIGHT' WAY

A German bio-psychologist, Onur Güntürkün, was curious whether the human tendency for right-sightedness (e.g., right-handed, right-footed, right-eyed), manifested itself in other situations as well. In trying to understand why human brains function asymmetrically, with each side controlling different abilities, he investigated whether kissing couples were more likely to lean their heads to the right than to the left[26]. He and his researchers observed 124 couples (estimated ages 13 to 70 years, not holding any other objects like luggage that might influence their behavior) in public places such as airports, train stations, beaches, and parks in the United States, Germany, and Turkey, of which 80 leaned their heads to the right when kissing.

In this activity, you will be exploring the following research question:

> What percentage of couples lean their heads to the right when kissing?

---

[26] Güntürkün, O. (2003). Human behaviour: Adult persistence of head-turning asymmetry. *Nature, 421*, 711.

DISCUSS THE FOLLOWING QUESTIONS.

1. Based on the data Güntürkün collected, provide a single number estimate to offer an answer to the research question.

2. Consider another study carried out using the same methodology, but using a **different sample of 124 couples**. Would you necessarily obtain the same answer to the research question (i.e., would the percentage of couples who lean their heads to the right when kissing be the same)? Explain why or why not.

3. Now imagine the same study were carried out 10 more times, **each time using a different set of 124 couples** from the same population. Make a conjecture about the percentage of couples from each of these studies who lean their heads to the right when kissing. Write these values down.

THE STANDARD ERROR

Remember, summary measures that describe a sample, like the mean and standard deviation, are called **statistics**. The standard deviation of a distribution of statistics (when each case is a statistic) is referred to as a **standard error**. The standard error is a quantification of the variation in a statistic due to random sampling or random assignment.

4. Enter your ten guesses for the percentage of couples from each of these studies who lean their heads to the right when kissing into a TinkerPlots™ case table and compute the standard deviation of these values. Record this value.

5. Is the value you just computed a standard deviation or a standard error? Explain.

6. What is the source of the variation in these values (i.e., what is the reason they are different)?

UNCERTAINTY OF THE ESTIMATE

Each of the imagined studies would be interested in answering the same research question, namely: What percentage of couples lean their heads to the right when kissing? Because they used different samples of couples, those studies might provide different estimates of the population percentage. For example, in the observed data (Güntürkün's data), 80 of the 124 couples in the sample, or 65%, leaned to the right when kissing. In another study the researchers might have found that 77 of the 124 couples, or 62% of the couples, leaned to the right when kissing.

When estimates like these are made from sample data, they often include an additional measure of the uncertainty of the estimate. This uncertainty is an acknowledgement of the fact that sample estimates will vary from sample to sample. To determine how much uncertainty is due to random sampling, the question that needs to be answered is: **How much variation is there in the different possible estimates when different random samples are used to make this estimate?**

This variability, as you learned in the previous activity, is quantified using the standard error.

MODELING THE VARIATION DUE TO RANDOM SAMPLING

Before you can compute a standard error for the estimate of the true proportion of couples leaning their heads to the right when kissing, you need to be able to draw many different samples of size 124. **A major obstacle is that you do not have access to the population.** You also do not have a model from which you can generate simulated data. (Note that if you had either of those two things, you would not need to estimate the proportion of couples leaning to the right when kissing, you could just determine what it was.)

What you have is the observed data. Without any other evidence as to what the true model is, the most informed choice is to **use the observed data as a stand-in**, or proxy, for the unknown model. This stand-in model can then be used to generate simulated data that represent many samples of 124 couples.

As you have witnessed throughout the course, a model generates many different samples of data. One major problem in substituting the observed data as the model from which you will generate simulated data is that the observed data values are fixed. Sampling from these values will generate the exact same values! This will not allow you to estimate the variation in the estimate across samples because the sample estimates will always be the exact same value!

In order to use the observed data as a model from which to generate simulated data without getting the exact same values, you need to sample from the observed data **with replacement**. This means that the same value can be sampled multiple times. The process of using the observed data as a stand-in for the unknown model and generating data by sampling with replacement is called **nonparametric bootstrapping**.

Nonparametric Bootstrapping Using TinkerPlots™

To carry out a nonparametric bootstrap analysis using TinkerPlots™, you:
- Set up a sampler in TinkerPlots™ based on the results in the observed data (80 out of 124 couples lean to the right) to generate simulated data for 124 couples.
- Plot the results for the trial.
- Collect the percentage of couples leaning to the right from the simulated data.
- Carry out 500 trials of the simulation.

Evaluating the Bootstrap Distribution

7. Plot the results from the 500 trials and sketch the plot below. Make sure to label the axis. Circle the mean and record its value.

8. Where is this distribution centered? Explain why it makes sense that the distribution is centered at this value.

9. Compute the standard error (use the `stDev()` function) based on this simulation.

10. What does this value represent (interpret its value)?

MARGIN OF ERROR

Consider the following poll reported in the New York Times on June 30, 2011:

> As the housing market slumped over the last few years with a speed and magnitude not seen since the Great Depression, aspects of homeownership have been debated as never before. There are tough questions about the role the government should take…includ[ing] how much of a down payment lenders should demand. Whether buyers need to come up with a 20 percent down payment—the standard for decades, but beyond the reach of many families now—is hotly debated. Fifty-eight percent of respondents say lenders should require this, while 36 percent say they should not. The nationwide telephone poll was conducted June 24–28 with 979 adults and has a margin of sampling error of plus or minus three percentage points for all adults.

In polls reported in the newspaper and online, the margin of error is almost always provided. For example, in the newspaper article above, the margin of error for the sample estimate (the percentage of all adults in the United States who believe that lenders should require a 20% down payment on a house) is given as ±3%. In studies reported in journals, however, the margin of error may not be reported. Luckily, the margin of error can easily be computed as,

Margin of Error = 2 x Standard Error

11. Using the standard error from the kissing study, compute the margin of error.

INTERVAL ESTIMATES

When statisticians report sample estimates, they provide the value of the estimate along with the quantification of the variation in the estimate expected from random sampling. As indicated previously, in popular media this is often reported as the **sample statistic and a margin of error**. For example, in the newspaper article above, the percentage of all adults in the United States who believe that lenders should require a 20% down payment on a house was reported as 58% ± 3%.

The same information can also be reported by adding and subtracting the value of the margin of error to the sample estimate and providing the actual **interval for the estimate**. For example, in the newspaper article above, the percentage of all adults in the United States who believe that lenders should require a 20% down payment on a house was reported as:

$$[55\%, 61\%].$$

Statisticians refer to this as an **interval estimate** because it gives an interval of plausible values for the percentage of all adults in the United States who believe that lenders should require a 20% down payment on a house. Based on the observed data, the best estimate for the "truth" is that 58% of all adults in the United States who believe that lenders should require a 20% down payment on a house. However, because of the uncertainty associated with random sampling, it may be that "truth" may be anywhere between 55% and 61%. All estimates in this range are just as believable.

12. Obtain the interval estimate for the true percentage of couples that lean to the right when kissing. Use the interval estimate to provide an answer to the research question.

MARGIN OF ERROR

To further help you understand ideas related to sampling error and the margin of error, we would like you to read a short pamphlet put together by the American Statistical Association's Section on Survey Research.

- Read Chapter 10 (*What is a Margin of Error?*) of the pamphlet, *What is a Survey?*. The pamphlet is available at http://www.amstat.org/sections/srms/pamphlet.pdf.

COURSE ACTIVITY: MEMORIZATION (PART II)

Recall the memory study that you analyzed to compare how many letters students in your class could recall when given the sequence grouped into familiar three-letter acronyms vs. unfamiliar groupings (with varying numbers of letters). The data from another class are presented below.

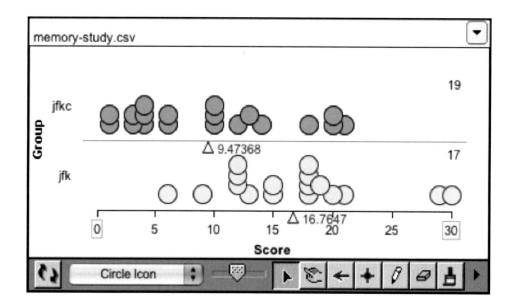

Data from the memorization study.

For this class's results, there was strong evidence against the null model of no difference between the two groups when the data was analyzed using a randomization test (p-value < 0.01). This suggested that the difference in the average number of letters memorized between the two treatments is not due to chance.

A natural follow-up question to rejecting the null model of no difference between the groups is:

> How many more letters, on average, can students remember when the words are grouped into familiar acronyms than when they are not?

To answer this question, you need to estimate **how much better one group is than the other**. This is called the **effect size**.

1. Use the observed data to find an estimate for the effect size. In other words, using the data, how many more letters, on average, can students remember when the words are grouped into familiar acronyms than when they are not?

EFFECT SIZE

Effect size is a term used to describe the extent to which sample results diverge from the expectations specified in the null hypothesis.

For example, consider a study that compares two weight loss programs. The null hypothesis might be that there is no difference in the average weight loss between the two programs. However, the researchers might find that in their samples of participants, the average weight loss is 30 pounds more for program A. In this case, 30 pounds is the effect size.

As another example, consider a study in which a tutoring program is examined. Using a sample of students, the researchers find that the students involved with the tutoring program have, on average, raised their school performance by one letter grade over a control group of students. This grade increase is the effect size of the tutoring program.

Quantifying the effect size helps researchers focus on whether hypothesis test results, even those showing strong statistical evidence against the null model, are meaningful or not. Reporting the size of the effect is considered good practice when presenting empirical research findings in many fields since it facilitates the interpretation of the substantive, as opposed to the statistical, significance of a research result.

In the two examples above, **the values offered for the effect size are estimates** of the true size of the effect. Using one set of subjects, for example, researchers might find the average difference in weight loss to be 30 pounds. However, with a different set of subjects, the size of the effect might be estimated to be 32 pounds.

Both are estimates of the true size of the effect, but may be different simply because of random assignment. Thus, as in the previous activity, it is important to also indicate a measurement of the uncertainty in the estimate.

Quantifying the Uncertainty of the Effect Size

In the previous unit, under the null model of no difference, the observations were **pooled** (put into one big group) and then the samples were drawn from the pooled sample. When this model is rejected (there is substantial statistical evidence against the null model), the conclusion is that the groups are different.

Under this conclusion, the strategy for modeling needs to change so that the simulation is conducted under the assumption of an alternative model. **It is now assumed that the samples represent two different groups with two different mean values.** Because it is believed that the samples represent two different groups with two different mean values, the replicate data sets are drawn from the two samples separately—the first replicate sample is bootstrapped only using the observations from the first sample, and the second replicate sample is bootstrapped only using the second observed sample.

Once you get the two replicate samples, you can compute the effect size—in the case of the weight loss example this would be the difference in means. Repeating this process a large number of times will give us a distribution of plausible differences in means for the population under the alternative model of a difference between the two groups. Using this distribution, you can calculate an estimate of the sampling error. The following steps illustrate this process:

- Consider our sample data as representative of the two group populations.
- Sample, with replacement, from each of the observed samples separately, matching the sample sizes used in the study.
- Compute the estimate of the effect size (e.g., the difference in means).
- Compute the uncertainty due to random assignment/samplng (i.e., the standard error).

USING TINKERPLOTS™ TO COMPUTE THE UNCERTAINTY IN THE EFFECT SIZE

Set up models in TinkerPlots™ that can be used to bootstrap the 17 memory scores for the familiar chunking group (JFK) and the 19 memory scores for the non-familiar chunking group (JFKC).

- Open the *Memorization.tp* file.
- Use TinkerPlots™ to carry out a single trial to bootstrap from each group separately (see instructions on next page).
- Create plots for each of the resulting bootstrapped samples.

2. Sketch the two plots below.

> **Bootstrapping from the Alternative Model**
> - Drag TWO `Samplers` from the object toolbar into the workspace.
> - Create an **empty** `Mixer` device in each of the sampler windows and set `Draw` to *1* for each sampler.
> - Copy and paste the 17 memory scores for the familiar chunking group (JFK) group into one of the mixers. Name this mixer *JFK_score*.
> - Copy and paste the 19 memory scores for the non-familiar chunking group (JFKC) group into the other mixer. Name this mixer *JFKC_score*.
> - Change the `Repeat` value for each sampler to the respective sample size for the group, and set the `Mixer` devices to sample `With Replacement`.
> - Click the `Run` button to simulate a single trial of the simulation.

3. Compute (do not collect yet) the effect size (i.e., the difference in the two mean scores) between the bootstrapped groups.

4. Run another trial for each simulation. Explain why you do not obtain the same difference in sample means every trial, even though you are again sampling 17 subjects from one group and 19 from the other group.

- Collect the sample means for each group sampler, separately.
- Run an additional 499 bootstrapped trials for each group sampler, separately.
- Create a table to calculate the effect size (difference in means) for the 500 bootstrapped trials (see instructions below).
- Plot the 500 differences.

Create a Table for the Effect Size (Difference in Means)

- Drag a `Case Table` from the object toolbar into the workspace.
- Create a new attribute named *mean_difference*.
- Open the `History of Results` for the first sampler. Highlight the *mean_JFK_score* attribute and select `Copy Attribute` from the `Edit` menu.
- In the newly created table, highlight the *mean_difference* column and select `Paste Attribute` from the `Edit` menu. This will paste the *mean_JFK_score* attribute in the column to the left of the *mean_difference* column.
- Open the `History of Results` for the second sampler. Highlight the *mean_JFKC_score* attribute and select `Copy Attribute` from the `Edit` menu.
- Again, highlight the *mean_difference* column in the table and select `Paste Attribute` from the `Edit` menu.
- Right-click the *mean_difference* attribute and use the formula editor to compute the difference between the *mean_JFK_score* and the *mean_JFKC_score* attributes.

5. Sketch a plot of the 500 effect sizes below.

6. Compute the mean value for the 500 bootstrapped effect sizes? Explain why the this distribution is centered at this value.

7. Compute the standard error for the effect size. Report that value here.

INTERVAL ESTIMATE FOR THE EFFECT SIZE

Remember the estimate of the standard error gives us an indication of how variable the estimates of the effect size will be from sample to sample. Often, you are also interested in using the sample estimate to indicate how large the size of the effect will be for a particular population. In other words, you are interested in using the sample estimate to indicate something about the unknown parameter.

In the last course activity, you used the sample estimate along with the estimate of the standard error to obtain an interval estimate. You can use the exact same method for obtaining an interval estimate for the effect size.

8. Compute the interval estimate for the effect size. Report that interval here.

9. Interpret the interval and use it to answer the research question.

COURSE ACTIVITY: WHY TWO STANDARD ERRORS

A common statistical question is how to estimate the mean of an unknown population when you cannot access the entire population. If you are able to obtain a random sample of data from such an unknown population, you can use the bootstrap method to estimate the mean of the population and the variability of sample means of random samples drawn from the population (i.e., the standard error). Using the original random sample as a model, you use the bootstrap resampling method to obtain many random samples of it.

> How can you obtain a precise estimate for a parameter using only a single sample?

EXPLORING RANDOM SAMPLES FROM AN UNKNOWN POPULATION

To address the question, you are going to draw random samples from an "unknown" population and examine the distribution of sample means computed from those samples.

- Open the *Population.tp* file.

A random sample of 15 values has been drawn for you from the "unknown" population. (To further convey the idea that the population is not known, it has been covered by a gray rectangle and marked with a question mark.) In addition, the sample mean has been collected into the History of Results table and plotted. The plot, called a **rug plot**, indicates the value of the sample mean by a short line segment rather than a circle.

- Collect 499 more random samples.
- Use the Divider tool to indicate where most of the means are in the distribution. To do this, think about where you might "cut off" the extreme values. It might help to look where the rug plot starts to "thin out" (i.e., noticeable space between the marks) at each end of the plot.

Remember that each line segment represents a mean value from a randomly drawn sample. Each of these sample means are, in turn, an estimate for the true population parameter, which is unknown to us. Since all of the means come from random samples drawn from the exact same population, **the variation seen in the rug plot is completely due to random sampling**. The divider gives us a sense of where most of the sample means are, and in turn, this helps to provide an idea of how much variability can be expected in the sample estimates just because of random sampling.

1. What are the values indicated by the divider tool endpoints?

2. What is the mean value for the 500 sample estimates?

3. How far above and below the mean value (approximately) are the divider tool endpoints?

4. What percentage of the sample means are contained in the middle part of your divider tool?

5. What would be your best guess of a good interval estimate for the mean of the unknown population based on where your divider tool endpoints are?

6. Compute the standard error of the sample means.

7. How many standard errors above (or below) the mean value are the divider tool endpoints?

8. How do you think this relates to why the margin of error is estimated as 2 x SE?

9. Compute the endpoints of an interval based on the mean of the 500 estimates ± 2 x SE. Roughly, what percentage of the collected sample means are inside the interval?

EXPLORING A SINGLE RANDOM SAMPLE

In practice, researchers do not draw several samples and thus, do not get to examine the characteristics of a distribution of sample estimates in order to obtain a measure for the standard error. Rather, the convention is to use a single sample to estimate the standard error. In this section, you will examine the bootstrap estimate for the standard error and interval estimate from a single sample to those estimates which were computed from re-sampling out of the "unknown" population.

- Open the provided sample of data corresponding to your group number.

10. Compute the mean for your sample.

11. Use the nonparametric bootstrapping method to obtain 500 bootstrapped means. Sketch a plot of the bootstrapped means below.

12. Find the estimate of the standard error based on the bootstrapped means and use it to calculate the margin of error.

13. Using the margin of error and the sample mean, compute an interval estimate for the true population mean.

14. What percentage of the collected bootstrapped means are inside the interval? (Hint: You may want to use the divider tool.)

Sketch your plot on the board. Be sure to label your x-axis and the lower and upper limits of your interval estimate. Also, record both the sample mean and the estimate for the standard error (based on your bootstrap distribution)

EXPLORING THE BOOTSTRAP DISTRIBUTION FROM DIFFERENT RANDOM SAMPLES

Examine the plots of your classmates. Each plot represents the distribution of 500 bootstrapped means. Remember, the bootstrapping for each plot is based on a different random sample of 15 values drawn from the original "unknown" population.

15. Which characteristics of the bootstrap distributions are the same? Which characteristics are different?

16. Explain why you would expect these similarities and differences.

EXPLORING THE BOOTSTRAP INTERVALS COMPUTED FROM DIFFERENT RANDOM SAMPLES

Now, consider the full computation for the interval estimate,

$$\text{Sample Estimate} \pm (2 \times SE)$$

To compute an interval estimate only requires two quantities, a sample estimate and an estimation of the standard error. Since the bootstrap allows researchers to estimate the standard error from a single sample, it is possible to compute an estimate of the uncertainty due to random sampling and interval estimate using only one sample!

Your instructor will enter the sample mean and estimate of the standard error from the different random sample into a table in TinkerPlots™ and display the interval estimates from all of the groups in a plot. Each interval estimate is represented by a horizontal line segment. The point in the middle of each line represents the sample mean. The plot also displays the mean of the "unknown" population using a vertical reference line.

17. Examine the estimates for the standard error (SE) in the table. Are they all the same? How close are they to each other? How similar are they to the estimate of the standard error you computed based on drawing 500 random samples from the "unknown" population (see Question 6)?

18. Examine the plot of the interval estimates. Are the endpoints for each interval estimate exactly the same? Are they close to each other? Are the widths of the intervals similar?

19. Considering how the interval estimate is computed, explain why you would expect these similarities and differences.

20. Examine the interval estimate **from your specific bootstrapped sample**. Is the population mean value inside your interval estimate?

21. Again, examine the plot of **all the interval estimates**. How many of the interval estimates include the population mean value inside the interval? How many of the interval estimates do not include the population mean value inside the interval?

WHAT DOES IT MEAN TO BE 95% CONFIDENT?

Interval estimates are sometimes referred to as **confidence intervals**. In interpreting these intervals, researcher will often say something like, "I am 95% confident that the population mean is between A and B." What do they really mean when they make a statement like this?

Not every interval estimate included the population mean. This can happen just because of chance (unlucky sampling). But what are your chances of producing a bootstrap interval estimate that includes the true population parameter (the population mean in this case)?

The figure below shows the bootstrap interval for each of 100 random samples drawn from the "unknown" population represented as a horizontal line segment. The sample mean is represented by a white mark in the middle of the interval. The vertical blue line represents the population mean.

To help you answer the questions below, the intervals have also been sorted in order of their sample means (the random sample having the lowest sample mean is on the bottom and the random sample having the highest sample mean is on top) and color-coded (if the interval includes the population mean value, it is drawn in grey, otherwise it is drawn in black).

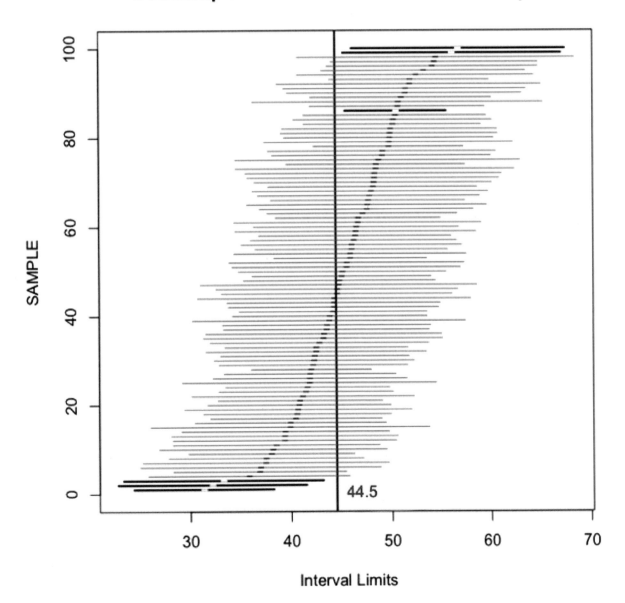

22. How many of the intervals include the population mean value (the grey intervals)? What percentage is this?

23. How many of the intervals do not include the population mean value (the black intervals)? What percentage is this?

24. Consider only the intervals that include the population mean value. What are the lowest and highest sample means for this set of intervals?

25. Compare the values from the last question to the endpoints of the values indicated by the divider tool endpoints you found in Question 1?

When you draw a random sample from an unknown population, you have a 95% chance of obtaining a sample that will produce an interval estimate based on the bootstrap that includes the population parameter. Because of this, you say that you are 95% confident that the true population parameter is in the interval estimate. The 95% refers to how often the method works over many samples...not how it performs for a single sample.

LEARNING GOALS: UNIT 3

The activities, homework, and reading that you have completed for the third part of this course have introduced you to several more fundamental ideas in the discipline of statistics. The ideas and concepts you were introduced to in this unit form the basis for statistical estimation. In addition, you have learned how to use TinkerPlots™ to carry out a bootstrap simulation.

Below are the key concepts and skills from Unit 3 that you should have learned.

LITERACY/UNDERSTANDING (TERMS AND CONCEPTS)

You should understand that:

- When describing or comparing distribution three features should be described/compared (shape, center, and variability)
- Standard deviation is often used to quantify the variation in a data set
- Standard error is used to quantify the uncertainty in an estimate that is due to random sampling or random assignment
- Interval estimates include a point estimate and margin of error
- Margin of error is typically computed as two standard errors
- Nonparametric bootstrapping can be used to estimate the standard error of a statistic (e.g., mean, percent/proportion, difference of means)
- Effect size is a quantification of how two groups differ
- When estimating the difference in means or proportions between two groups, the simulation is carried out assuming that the alternative hypothesis of group differences *is* true (i.e., the data are not pooled together, instead the data from each group is used independently to generate bootstrap samples)

SELECTING/USING MODELS

You should be able to use TinkerPlots™ to:

- Generate simulated data from a model that includes two separate sampling devices to produce the bootstrap distribution of the effect size under the assumption of an effect of treatment or group differences (distributions are centered at the sample effect)

TINKERPLOTS™ SKILLS

You should also be able to do the following using TinkerPlots™:

- Compute the standard error of a statistic

In the next course activity, *Unit 3 Wrap-Up & Review*, you will have a chance to assess yourself on whether or not you have mastered these ideas through a variety of practice and extension problems. As a pre-cursor to this activity, you may want to review the readings and activities in Unit 3.

COURSE ACTIVITY: UNIT 3 WRAP-UP & REVIEW

TERMINOLOGY FOR UNIT 3

At this point, you should be familiar with the following terms. Write down what each term represents as well as any notes that may help you remember.

1. Shape

2. Center

3. Variation

4. Standard Deviation

5. Standard Error/Precision of the Estimate

6. Margin of Error

7. Interval Estimate

8. Effect Size

RELATIONSHIP BETWEEN P-VALUE AND STANDARD ERROR

The difference in mean arrival delay times for two airlines Mesa and American Eagle, was calculated by taking the mean arrival delay time for Mesa Airlines and subtracting it from the mean arrival delay time for American Eagle Airlines. This difference was calculated for two cities: Madison, WI (−15.9; American Eagle arrived, on average, 15.9 minutes earlier than Mesa), and Fort Wayne, TX (−20; American Eagle arrived, on average 20 minutes earlier than Mesa).

Use the *Airlines-Variability.tp* file to help you answer the following questions.

9. Use TinkerPlots™ to calculate the standard deviation of the arrival delay times for Madison, WI. Do the same for Fort Wayne, TX. Report the values of each standard deviation.

10. For the Madison, WI data, use TinkerPlots™ to carry out a bootstrap analysis with 500 trials to estimate the standard error for the difference in means. Report this value.

11. For the Fort Wayne, TX data, use TinkerPlots™ to carry out a bootstrap analysis with 500 trials to estimate the standard error for the difference in means. Report this value.

12. Which of the cities has the larger standard error?

A randomization test was carried out to examine the the null hypothesis of no difference between the two mean arrival times for each city. The p-value for the analysis of the Madison, WI data was 0.132 whereas the p-value for the analysis of the Fort Wayne, TX data was 0.226.

13. Explain why you could predict that the p-value for Madison, WI would be smaller than the p-value for Fort Wayne, TX, even though Fort Wayne, TX has the larger observed difference in means, by considering the two standard errors you computed previously,

BALSA WOOD

Student researchers investigated whether balsa wood is less elastic after it has been immersed in water[27]. They took 44 pieces of balsa wood and randomly assigned half to be immersed in water and the other half not to be immersed in water. They measured the elasticity by seeing how far (in inches) the piece of wood would project a dime into the air. The observed difference in mean elasticities between those pieces of balsa wood immersed in water and those not immersed in water was 4.16 inches. A bootstrap analysis found that the standard error for the difference in means was 0.70.

14. Compute the 95% interval estimate for the effect size.

15. Interpret the interval.

16. Based on the interval estimate, does this study provide plausible evidence to suggest that balsa wood immersed in water has a higher mean elasticity than balsa wood that is not immersed in water? Explain.

---

[27] Rossman, A. J., Chance, B. L., & Lock, R. H., (2009). *Workshop statistics: Discovery with data and Fathom* (3rd ed.). Emeryville, CA: Key College Publishing.

MICROSORT™

The *Genetics and IVF Institute* (http://www.microsort.com/) is currently studying methods to change the odds of having a girl or boy. MicroSort™ is a method used to sort sperm with X- and Y-chromosomes. The method is currently going through clinical trials. Women who plan to get pregnant and prefer to have a girl can go through a process called X-Sort™. As of 2008, 945 women have participated and 879 of those women have given birth to girls.

17. Compute the 95% interval estimate of the percentage of female births for women that undergo X-Sort™.

18. Interpret the interval.

19. Why do you set the Repeat value in your simulation equal to the sample size?

20. Suppose more data has been collected since 2008. If the number of women had increased to 3,000 but the observed percentage of female births remained the same, what would you expect to happen to the size of your interval?

21. Test out your conjecture by creating a new interval using a sample size of 3,000. Report your new interval estimate. Was your expectation in the previous question correct?

22. How many trials did you run in your simulations?

23. What is the difference between sample size and number of trials?

MARIJUANA

Pope and Yurgelun-Todd studied whether frequent marijuana use is associated with residual neuropsychological effects[28]. College undergraduate students were recruited to participate in the study. A total of 65 heavy marijuana users and 64 light users participated. The students took a neuropsychological test which involves sorting cards. The average number of cards correctly sorted for the heavy marijuana users was 51.3 and for the light marijuana users was 53.3.

Use the *Marijuana.tp* file to help you answer the following questions.

24. Carry out a randomization test to determine whether heavy marijuana users sort fewer cards correctly, on average, compared to light marijuana users? Report the pertinent results below.

25. Provide an interval estimate for the effect size. Report the pertinent results and interpret the interval below.

---

[28] Pope, H. G., & Yurgelun-Todd, D. (1996). The residual cognitive effects of heavy marijuana use in college students. *Journal of the American Medical Association, 274*(7). 521–527.

Made in the USA
San Bernardino, CA
13 January 2017